T0299352

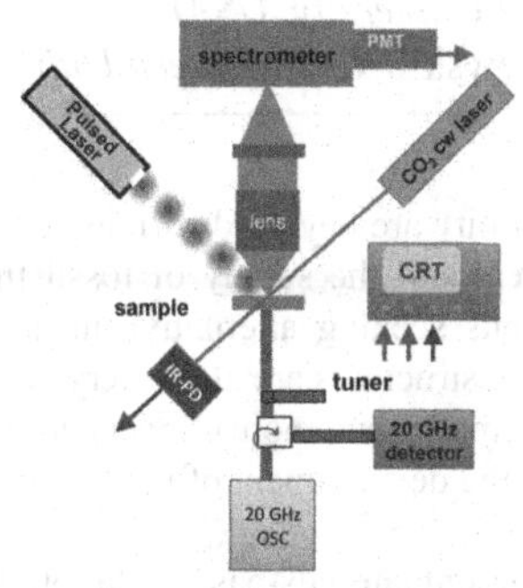

THEORY AND METHODS OF PHOTOVOLTAIC MATERIAL CHARACTERIZATION

Optical and Electrical Measurement Techniques

World Scientific Series in Materials and Energy

ISSN: 2335-6596

Energy and sustainability are keywords driving current science and technology. Concerns about the environment and the supply of fossil fuel have driven researchers to explore technological solutions seeking alternative means of energy supply and storage. New materials and material structures are at the very core of this research endeavor. The search for cleaner, cheaper, smaller and more efficient energy technologies is intimately connected to the discovery and the development of new materials.

This collection focuses onmaterials-based solutions to the energy problem through a series of case studies illustrating advances in energy-related materials research. The research studies employ creativity, discovery, rationale design and improvement of the physical and chemical properties of materials leading to new paradigms for competitive energy-production. The challenge tests both our fundamental understanding of material and our ability to manipulate and reconfigure materials into practical and useful configurations. Invariably these materials issues arise at the nano-scale!

For electricity generation, dramatic breakthroughs are taking place in the fields of solar cells and fuel cells, the former giving rise to entirely new classes of semiconductors; the latter testing our knowledge of the behavior of ionic transport through a solid medium. Inenergy-storage exciting developments are emerging from the fields of rechargeable batteries and hydrogen storage. On the horizon are breakthroughs in thermoelectrics, high temperature superconductivity, and power generation. Still to emerge are the harnessing of systems that mimic nature, ranging from fusion, as in the sun, to photosynthesis, nature's photovoltaic. All of these approaches represent a body of materials–based research employing the most sophisticated experimental and theoretical techniques dedicated to a commongoal. The aim of this series is to capture these advances, through a collection of volumes authored by leading physicists, chemists, biologists and engineers that represent the forefront of energy-related materials research.

Published

Vol. 13 *Theory and Methods of Photovoltaic Material Characterization: Optical and Electrical Measurement Techniques*
by Richard K. Ahrenkiel (Colorado School of Mines, USA and National Renewable Energy Laboratory, USA) and S. Phil Ahrenkiel (South Dakota School of Mines & Technology, USA)

For further details, please visit: http://www.worldscientific.com/series/mae

(Continued at the end of the book)

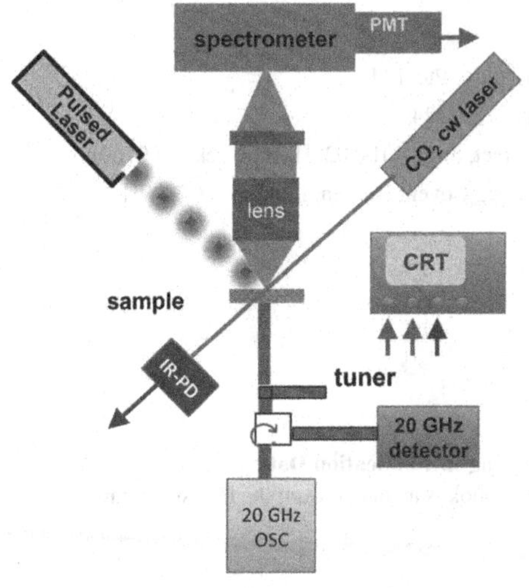

MATERIALS AND ENERGY – Vol.13

THEORY AND METHODS OF PHOTOVOLTAIC MATERIAL CHARACTERIZATION

Optical and Electrical Measurement Techniques

Richard K. Ahrenkiel
Colorado School of Mines, USA
National Renewable Energy Laboratory, USA

S. Phillip Ahrenkiel
South Dakota School of Mines & Technology, USA

World Scientific

NEW JERSEY • LONDON • SINGAPORE • BEIJING • SHANGHAI • HONG KONG • TAIPEI • CHENNAI • TOKYO

Published by

World Scientific Publishing Co. Pte. Ltd.
5 Toh Tuck Link, Singapore 596224
USA office: 27 Warren Street, Suite 401-402, Hackensack, NJ 07601
UK office: 57 Shelton Street, Covent Garden, London WC2H 9HE

British Library Cataloguing-in-Publication Data
A catalogue record for this book is available from the British Library.

Materials and Energy — Vol. 13
THEORY AND METHODS OF PHOTOVOLTAIC MATERIAL CHARACTERIZATION
Optical and Electrical Measurement Techniques

ISBN 978-981-3277-09-0

For any available supplementary material, please visit
https://www.worldscientific.com/worldscibooks/10.1142/11179#t=suppl

Desk Editor: Rhaimie Wahap

Typeset by Stallion Press
Email: enquiries@stallionpress.com

Printed in Singapore

To my grandchildren; Katherine, Christian, Luke, Benjamin, and Grace and all those of their generation. May the climate of planet Earth remain as friendly to you as it has been to my generation.

Richard K. Ahrenkiel

Preface

This textbook will not attempt to cover the vast field of semiconductor characterization. Such information can be found in books such as *Semiconductor Materials and Device Characterization* written by my late colleague Prof. Dieter Schroder at Arizona State University. Instead this text will cover in detail some of the more specialized techniques that came out of the current authors experience in photovoltaic characterization.

For the first author (RKA), this textbook arises from the 37 years of association with the Measurement and Characterization activities of the National Renewable Energy Laboratory (NREL) in Golden, Colorado. I joined the Measurements and Characterization (M&C) Branch of the (then) Solar Energy Research Institute (SERI) in 1981. My first task, as defined by Branch Chief Dr. Lawrence Kazmerski, was to develop electrical and optical techniques to support the research and development activities of the small in-house (at that time) research groups. In addition, the Branch supported a much larger group of academic and industrial subcontractors that were charged with the invention and development of new and innovative photovoltaic technologies by the U.S. Department of Energy. The M&C branch provided a characterization service for the DOE subcontract program, thereby avoiding a duplication of infrastructure at each external institution. At that time, the M&C Branch had a basic infrastructure of the commercially available diagnostics based on vacuum based instruments such as electron microscopes (TEM and SEM), Auger, secondary ion mass spectroscopy (SIMS), and X-ray photoemission spectroscopy (XPS).

One of the first requests, from my materials growth partners, was for the measurements of minority carrier lifetime in new materials. At that time, the only on-site technique was a four-probe setup with flashlamp excitation which could be used to measure photoconductive decay in silicon wafers of ingots. The laboratory materials under intense investigation at that time were thin-films of III-V and II-IV semiconductors such as GaAs, CdTe and copper indium selenide (CIS). Consequently, my research partner, master technician Donald Dunlavy built the first (at NREL) time-resolved photoluminescence system. This technique has proven to be a standard method for evaluating direct bandgap thin films over the years.

A second challenge arose out of the work in the early 1990s based on using small (1 mm diameter) spherical silicon particles as the building blocks for inexpensive solar cells. Contacting techniques would not function in this case, so a contactless radio frequency technique evolved based on the early results of the Bell Laboratory workers. This evolved into the resonant-coupled photoconductive decay technique (RCPCD) when it was discovered that large increases in sensitivity could be achieved if the detection system was engineered to be a very high quality-factor (Q) resonant circuit. The RCPCD technique became extremely useful when the materials research extended into the low-bandgap III-V materials used in the thermo-photovoltaic (TPV) program. These materials could not be characterized at that time owing to the lack of photon counting detectors that were sensitive in the 1- to 3-μm wavelength range.

As the need and value of advanced characterization increased, numerous other techniques evolved at NREL and elsewhere that will be described in this manuscript. The NREL program at that time was labeled technique development. This additional task was to develop new and better techniques that accelerate the development of new materials.

The format of this text is to present the measurement theory of a given technique. The theory is followed by data that represents typical experiment results that are obtained from the measurement technique. Much of the data and figures were generated by the authors and their colleagues at NREL and collaborating institutions as part of the ongoing research at the various sites. Over that time period, the first author and colleagues recorded over ten thousand measurements on almost every known semiconductor material. These materials ranged from thin films to large ingots, and from silicon-based structures and devices to carbon nanotubes. The goal here is to collect together a summary of this multi-decade study.

There is a vast body of research on other techniques at institutions around the globe and reported in a large number of technical journals. There is no attempt to cover that large body of work here.

The hope of the authors is that this manuscript will aid ongoing work at world-wide research and development institutions on photovoltaic and electronic materials. We hope that this manuscript will aid students and researchers, who work in the photovoltaic and other semiconductor technologies, with new tools.

Richard K. Ahrenkiel
Colorado School of Mines
and
National Renewable Energy Laboratory

S. Phillip Ahrenkiel
South Dakota School of Mines and Technology

August 2018

Acknowledgements

The first author (RKA) has many colleagues and collaborators to thank for providing the key contents of this book. First, and foremost, I would like to thank Dr. Steve Johnston and Mr. Donald Dunlavy for producing many of the innovative ideas and much of the effort that underlies the innovative technique development described in this manuscript. I also wish to thank Ms. Tami Sandberg for the invaluable help in the literature search that underlies this work. Also, the unique graphics skills of Mr. Al Hicks made the figures and illustrations in this book possible. Over, the 37 years of association with NREL and the 28 years with the Colorado School of Mines, I have enjoyed many invaluable collaborations and associations. These include Drs. Dean Levi, Brian Keyes, Randy Ellingson, Wyatt Metzger, Mark Wanlass, John Moseley, Ari Feldman, Brian Simonds, Darius Kuciauskas, Tim Coutts, Gregory Horner, Arthur Nozik, and Ms. Pat Dippo. Finally, I would like to thank Dr. Lawrence Kazmerski for providing the kind of unique research climate that allowed innovation to flourish during his tenure as the Director of Measurement and Characterization at NREL.

Contents

CHAPTER 1

Semiconductor Fundamentals and Background

1.1. Tetrahedral Semiconductors

The material most widely associated with photovoltaics is silicon (Si), and its well-established properties are often the reference from which those of other PV materials are drawn. Silicon is a group-IV, tetravalent metalloid, inert, grey solid, with a high melting point (1400°C), a chemical analog to carbon. The four valence electrons of both Si and carbon (C) make these elements compatible with tetrahedral coordination. Building from the core of a neon (Ne) atom, a pair of spin-opposite electrons are first added to the 3s level, followed by two electrons in the 3p levels, which are spin aligned, according to Hund's rule. The energy levels are shown in Fig. 1.1. Hybridization by linear combination of the 3s and 3p atomic orbitals can be used to form four molecular orbitals with minimal overlap. Whereas the hybridized levels have higher energy for isolated atoms, they result in an overall reduction in energy in the solid by forming tetrahedral bonds.

The molecular-orbital wave functions are constructed from linear combinations of these four, equally weighted orbitals:

$$\psi = \frac{1}{2}\psi_s \pm \frac{1}{2}\psi_{p_x} \pm \frac{1}{2}\psi_{p_y} \pm \frac{1}{2}\psi_{p_z} \tag{1.1}$$

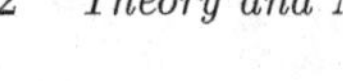

Fig. 1.1. $3sp^3$ hybridization.

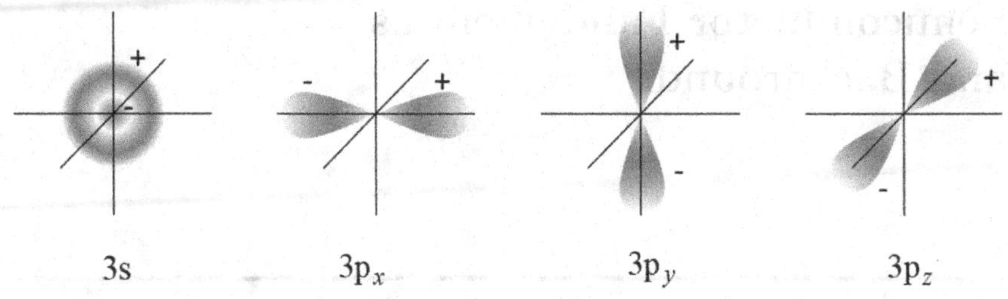

Fig. 1.2. Atomic orbitals that contribute to $3sp^3$ molecular orbitals.

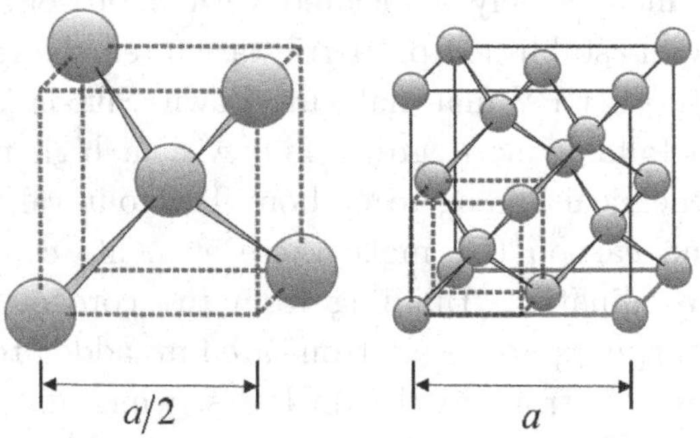

Fig. 1.3. The tetrahedron formed of nearest neighbors in Si can be inscribed within a cube of dimension $a/2$, where a is the cubic lattice parameter.

From the eight distinct possibilities generated by the various sign combinations, four can be selected with a tetrahedral arrangement. These functions are illustrated in Fig. 1.2.

Pure Si has the diamond crystal structure, most associated with carbon, which is comprised of two interpenetrating face-centered cubic lattices, both occupied by Si atoms, but separated by 1/4 the cube diagonal. The structures are shown in Fig. 1.3. Tetrahedral bonds associate each atom in either sublattice to the four nearest neighbors in the other sublattice. The tetrahedron formed among each atom and its four nearest neighbors can be inscribed within a cube of dimension $a/2$. Four such tetrahedra arranged with edges touching can be assembled into the full cubic unit cell.

The electronic properties of crystals are influenced by the available states supported by the material and the distribution of electrons within these states. Continuous bands form that dictate the variation of energy with wave vector. The occupation of these bands depends on the numbers of electrons and temperature. Thus, the electrical-conduction properties of materials are closely linked to the band structure, carrier density, and temperature. Of utmost importance is the bandgap E_g — the energy between the highest occupied state and the lowest unoccupied state, which are derived from the highest-occupied and lowest-unoccupied molecular orbitals, [HOMO and LUMO, respectively] of isolated molecules. A schematic of these energy levels are shown in Fig. 1.4.

Semiconductor bandgaps are either direct or indirect, which refers to the alignment in crystal momentum space of the VB maximum and the CB minimum. For a direct-gap semiconductor, these are aligned in wave vector; for an indirect-gap semiconductor they are not aligned. The implication is that no change in crystal momentum is required for electronic transitions between the VB

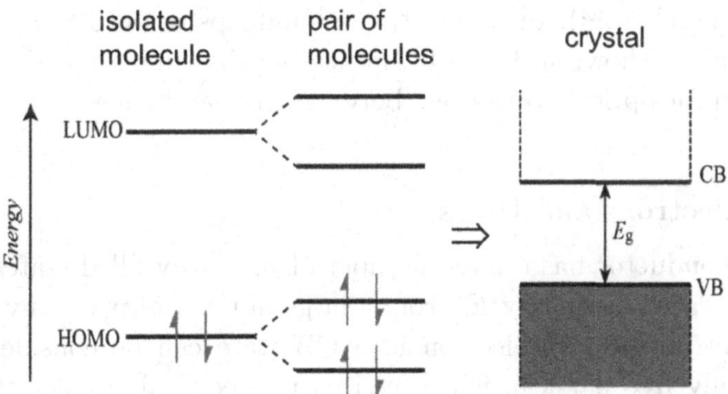

Fig. 1.4. The semiconductor bandgap can be related back to the HOMO-LUMO energy difference of isolated molecules.

and CB. Thus, the transition may represent only a change in energy of an electron. That change in energy can occur by the absorption or emission of a photon which, as a massless particle, has negligible momentum. An indirect transition, on the other hand, requires concomitant momentum change, generally by the emission or absorption of a phonon.

Bandgap is only one parameter describing the rather complex electronic structure of semiconductors. For example, the electronic bands of a three-dimensional crystal potential will not be spatially isotropic. The conductive properties of a semiconductor are dramatically influenced by doping, temperature, and illumination. These variations are exploited to enable control of electrical conduction for numerous electronic applications.

Advanced theoretical models have been developed for first-principles computation of semiconductor band structures, but the general form of the bands can often be demonstrated from relatively simple methods, involving interactions among valence electrons of nearest-neighbor atoms. Vogl and coworkers[1] have described one such semi-empirical tight-binding method that uses a small set of basis functions and provides data for all of the common tetrahedral semiconductors. Relativistic (spin-orbit) corrections can be incorporated with only small modifications, as outlined by Datta,[2] which remove degeneracies among the VB states. The direct bandgap of GaAs corresponds to the alignment VB and CB edges at the Brillouin zone center (labeled $\mathscr{P}$). In contrast, the bandgaps of both Si and AlAs are indirect (shown in Fig. 1.5), indicating that phonon assistance is required for optical transitions between the band edges.

1.2. Electrons and Holes

A semiconductor has a large number of primarily filled states in its VB separated by energy E_g from a large number of primarily empty states in its CB. An electron in a CB state can be considered an essentially free particle, which can carry electrical current through the crystal by virtue of its velocity and charge. Correspondingly, the absence an electron (negative charge) in the allows an imbalance

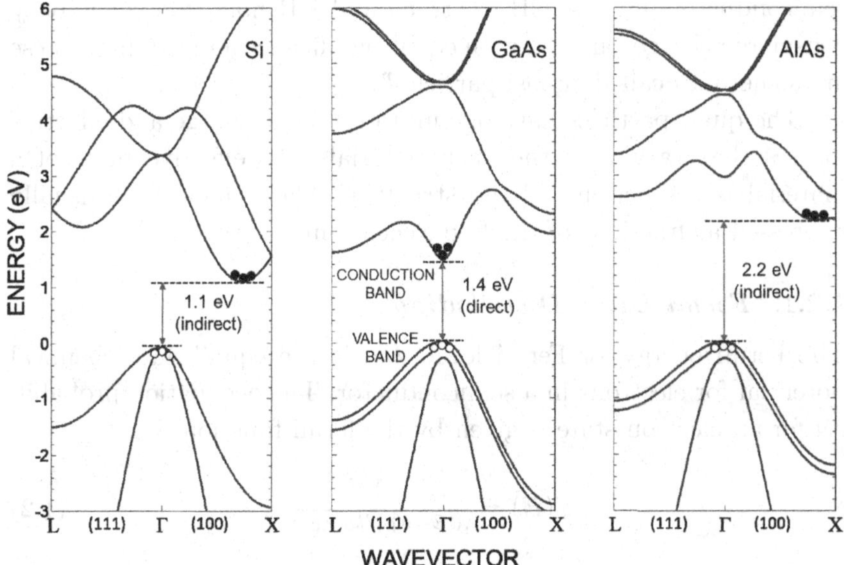

Fig. 1.5. Electronic band structures of common semiconductors. Calculations were performed using the method described by Datta.[2]

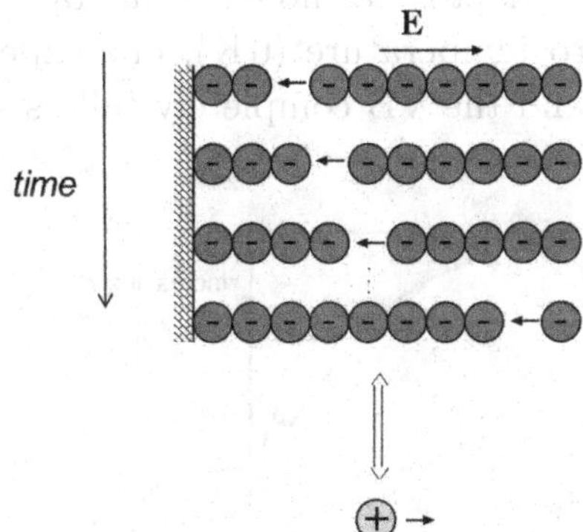

Fig. 1.6. Schematic representation of hole motion correspondence to composite electron motion.

in the net flow of charge in some direction (left). This process is illustrated in Fig. 1.6.

This motion is equivalent to the motion of a hole (positive charge) in the opposite direction (right). In the nomenclature of semiconductor physics, CB electrons and VB holes are collectively considered charge carriers, with equal significance and validity. These are sometimes called "quasi-particles".

The quasi-particles have one unit of electric charge and effective masses that vary with the host material. The effective mass of a material is a function of band structure. These masses are usually expressed as fractions of the free electron mass.

1.2.1. *Fermi-Dirac Distribution*

The Fermi energy, or Fermi level, E_{F}, is the equilibrium chemical potential for electrons in a semiconductor. The occupation probability for an electron state is given by the fermi function:

$$f(E) = \frac{1}{e^{(E-E_{\mathrm{F}})/kT} + 1} \tag{1.2}$$

(alternatively written $f(E - E_{\mathrm{F}})$, or $f[(E - E_{\mathrm{F}})/kT]$). The probability a state of energy E contains an electron is precisely 50% when $E = E_{\mathrm{F}}$. However, E_{F} often lies within the bandgap, in the range at which a pure semiconductor has no electron states.

At absolute zero temperature (0 K), one expects the CB to be completely empty and the VB completely full as shown in Fig. 1.7.

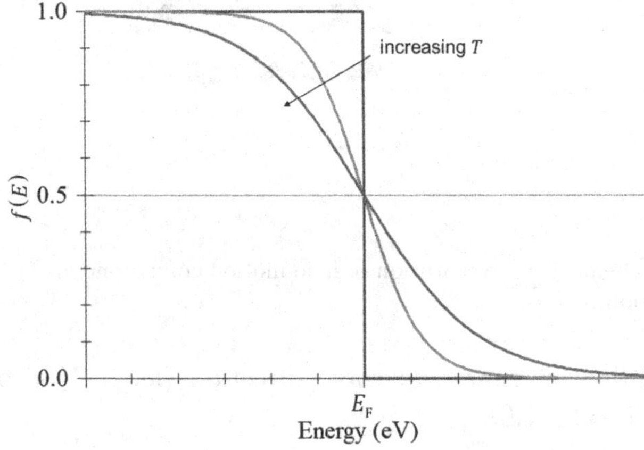

Fig. 1.7. Fermi-Dirac distribution function at various temperatures.

But at finite temperatures, the probability of thermally generated electrons in the CB and holes in the VB is non-zero.

1.2.2. *Carrier Concentrations*

The number of carriers per unit volume, per unit energy, depends on the electronic density of states (DOS) per unit volume of the appropriate band. The number of states in a 3-D box with linear dimension L having wavenumber in the range from 0 to k, with two possible spin states, is found to be:

$$N(k) = \frac{8}{3}\pi k^3 L^3 \tag{1.3}$$

Based on a qualitative understanding of the band structure, assuming the semiconductor has a direct gap at the gamma point ($k = 0$), we can write the dispersion relation near each band edge as

$$E_{\mathrm{CB}}(k) \approx E_{\mathrm{C}} + \frac{h^2 k^2}{2m_{\mathrm{C}}}, \quad (E \geq E_{\mathrm{C}}) \tag{1.4}$$

$$E_{\mathrm{VB}}(k) \approx E_{\mathrm{V}} - \frac{h^2 k^2}{2m_{\mathrm{V}}}, \quad (E \leq E_{\mathrm{V}}) \tag{1.5}$$

The denominators contain the CB and VB effective masses, which are inversely proportion to the band curvatures:

$$\frac{1}{m_{\mathrm{C}}} = \frac{1}{h^2}\frac{d^2}{dk^2}E_{\mathrm{CB}}(k)\bigg|_{k=0} \tag{1.6}$$

$$\frac{1}{m_{\mathrm{V}}} = -\frac{1}{h^2}\frac{d^2}{dk^2}E_{\mathrm{VB}}(k)\bigg|_{k=0} \tag{1.7}$$

Changing to an energy representation, the number of states above the CB band edge with energy $E \geq E_{\mathrm{C}}$ is given by:

$$N_{\mathrm{C}}(E) = \frac{8(2m_{\mathrm{C}})^{3/2}\pi L^3}{3h^3} \cdot (E - E_{\mathrm{C}})^{3/2}, \quad (E \geq E_{\mathrm{C}}) \tag{1.8}$$

The same approach is applied to the VB to the find the number of states $E \leq E_{\mathrm{V}}$.

$$N_{\mathrm{V}}(E) = \frac{8(2m_{\mathrm{V}})^{3/2}\pi L^3}{3h^3} \cdot (E_{\mathrm{V}} - E)^{3/2}, \quad (E \leq E_{\mathrm{V}}) \tag{1.9}$$

The densities of states per unit volume in the CB and VB are:

$$g_{\mathrm{C}}(E) = \frac{1}{L^3} \cdot \frac{dN_{\mathrm{C}}(E)}{dE} = 4\pi \left(\frac{2m_{\mathrm{C}}}{h^2}\right)^{3/2} \cdot (E - E_{\mathrm{C}})^{1/2}, \quad (E \geq E_{\mathrm{C}}) \tag{1.10}$$

$$g_{\mathrm{V}}(E) = \frac{1}{L^3} \cdot \frac{dN_{\mathrm{V}}(E)}{dE} = 4\pi \left(\frac{2m_{\mathrm{V}}}{h^2}\right)^{3/2} \cdot (E - E_{\mathrm{V}})^{1/2}, \quad (E \leq E_{\mathrm{V}}) \tag{1.11}$$

This dispersion relationship is plotted in Fig. 1.8. The concentrations per unit energy of electrons in the CB and holes in the VB are then:

$$\frac{dn}{dE} = g_{\mathrm{C}}(E) \cdot f(E - E_{\mathrm{F}}) \tag{1.12}$$

$$\frac{dp}{dE} = g_{\mathrm{V}}(E) \cdot [1 - f(E - E_{\mathrm{F}})] \tag{1.13}$$

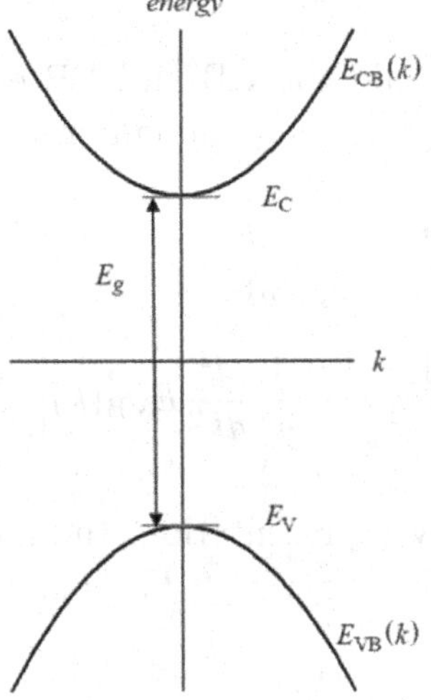

Fig. 1.8. Dispersion within parabolic bands of a direct-bandgap semiconductor.

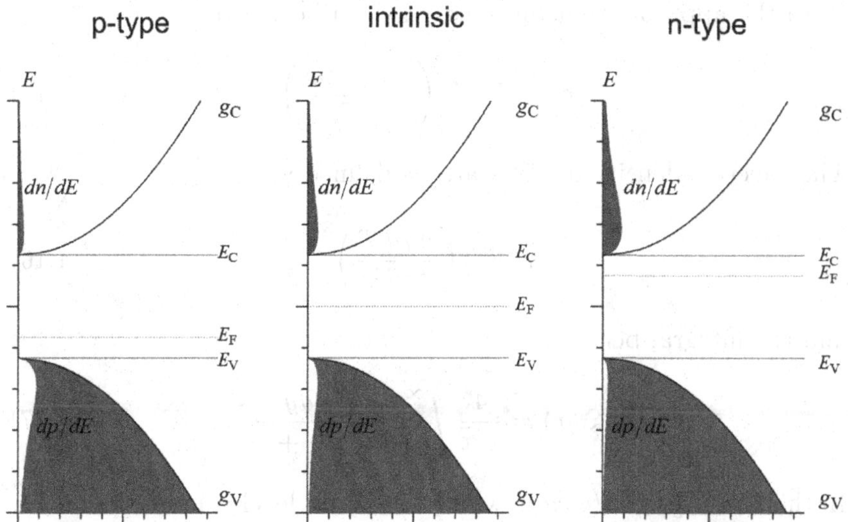

Fig. 1.9. Depiction of filled electronic states in p-type, intrinsic, and n-type semiconductors. The fraction of filled states in the CB and VB are shaded. (A high temperature is assumed for clarity.)

Within the parabolic-band approximation, we can plot these concentrations as functions of energy for different values of E_F as shown in Fig. 1.9. For a pure semiconductor, n and p are equal; the material is called intrinsic, and the corresponding position E_F is the intrinsic level E_i. A shift up (down) of E_F corresponds to an increase (a decrease) in n and a decrease (an increase) in p. The material is called n-type or p-type, depending on the majority-carrier type.

We are often unconcerned with the details of the carrier concentrations; we would like to know the total concentration of electrons in the CB and holes in the VB, regardless of energy. In the case of the CB,

$$\begin{aligned} n &= \int_{E=E_C}^{\infty} \frac{dn}{dE} \cdot dE \\ &= 4\pi \left(\frac{2m_C}{h^2}\right)^{3/2} \int_{E=E_C}^{\infty} (E - E_C)^{1/2} \cdot \frac{1}{e^{(E-E_F)/kT} + 1} \cdot dE \end{aligned} \tag{1.14}$$

With the appropriate definitions, this can be written as

$$n = N_C \cdot \Im\left(\frac{E_C - E_F}{kT}\right) \tag{1.15}$$

The effective density of CB states is defined as:

$$N_C = 2\left(\frac{m_C kT}{2\pi\hbar^2}\right)^{3/2} \tag{1.16}$$

and the integral becomes

$$\Im(x) = \frac{4}{\sqrt{\pi}} \int_{y=0}^{\infty} \frac{y^2 \cdot dy}{e^x \cdot e^{y^2} + 1} \tag{1.17}$$

In the *non-degenerate* case, where the Fermi level is more than a few kT below the CB band edge, the integral $\Im(x)$ reduces to

$$\Im(x) \approx e^{-x} \tag{1.18}$$

This allows a simple expression for the total concentration of CB electrons:

$$n = N_C \cdot e^{-(E_C - E_F)/kT} \tag{1.19}$$

The effective density of states concept allows representation of the CB as a single energy level at the band edge with a multiplicity of N_C states.

Likewise, for holes

$$p = N_V \cdot \Im\left(\frac{E_F - E_V}{kT}\right) \approx N_V \cdot e^{-(E_F - E_V)/kT} \tag{1.20}$$

using the defined effective density of VB states:

$$N_V = 2\left(\frac{m_V kT}{2\pi\hbar^2}\right)^{3/2} \tag{1.21}$$

The distribution functions using Boltzmann and Fermi functions are shown in Fig. 1.10. It is useful to define a reference constant for

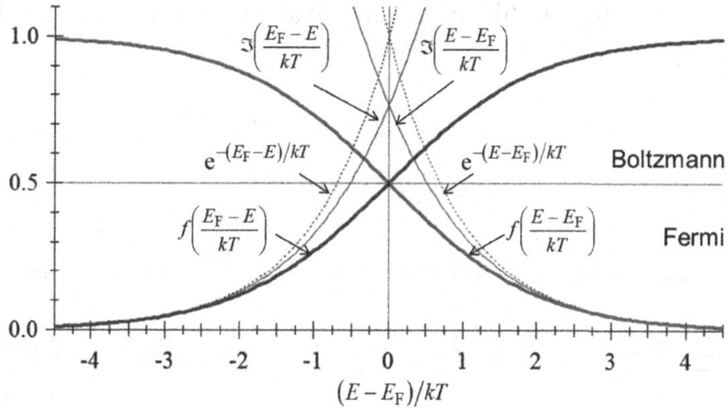

Fig. 1.10. Thermal distribution functions discussed.

the effective DOS at room temperature ($kT = 0.026\,\mathrm{eV}$)

$$N_0 = 2\left(\frac{m_0 kT}{2\pi\hbar^2}\right)^{3/2} = 2.5 \times 10^{19}\,\mathrm{cm}^{-3} \tag{1.22}$$

Then

$$N_C = \left(\frac{m_C}{m_0}\right)^{3/2} \cdot N_0 \tag{1.23}$$

$$N_V = \left(\frac{m_V}{m_0}\right)^{3/2} \cdot N_0 \tag{1.24}$$

1.2.3. *The Law of Mass Action*

In the non-degenerate case, the $n \cdot p$ product is a constant for a given semiconductor in equilibrium, regardless of the individual concentrations:

$$n \cdot p = N_C \cdot N_V \cdot \mathrm{e}^{-(E_C - E_V)/kT} = N_C \cdot N_V \cdot \mathrm{e}^{-E_g/kT} = n_i^2 \tag{1.25}$$

where

$$n_i = \sqrt{N_C \cdot N_V} \cdot \mathrm{e}^{-E_g/2kT} \tag{1.26}$$

The quantity n_i is called the *intrinsic* carrier concentration, which is the concentration of CB electrons and VB holes of the pure semiconductor in equilibrium. This leads to the alternative forms for n and p:

$$n = n_i \cdot \mathrm{e}^{(E_\mathrm{F}-E_i)/kT} \tag{1.27}$$

$$p = n_i \cdot \mathrm{e}^{(E_i-E_\mathrm{F})/kT} \tag{1.28}$$

The intrinsic energy, E_i, is the position of the Fermi level when $n = p$:

$$E_i = \frac{E_\mathrm{C}+E_\mathrm{V}}{2} - \frac{kT}{2} \cdot \ln\left(\frac{N_\mathrm{C}}{N_\mathrm{V}}\right) = \frac{E_\mathrm{C}+E_\mathrm{V}}{2} - \frac{3kT}{4} \cdot \ln\left(\frac{m_\mathrm{C}}{m_\mathrm{V}}\right) \tag{1.29}$$

If $m_\mathrm{C} = m_\mathrm{V}$, then E_i is exactly in the center of the gap. It is easily verified that, if $E_\mathrm{F} = E_i$, then $n = p = n_i$ (i.e., the material is intrinsic). From the above, the intrinsic level in silicon E_i is -5.57 meV below the midgap level.

By using the band-edge energies for reference, we can draw simplified energy diagrams showing energy vs. position, rather than momentum. The energy vs. position representation will be used almost exclusively in our discussion. It should be noted that the CB electrons are not truly free; they remain bound to the solid to maintain net charge neutrality. The vacuum level, E_vac, is defined as the minimum energy for an electron that is unbound from the solid. These levels are shown schematically in Fig. 1.11.

The work function, $\Phi_w = E_\mathrm{vac} - E_\mathrm{F}$, is the energy to free an electron with energy E_F from the solid. This is a useful parameter for metals, but for semiconductors, E_F lies with the bandgap, where we expect the density of states to be zero. A more useful quantity for semiconductors is the electron affinity, $\chi = E_\mathrm{vac} - E_\mathrm{C}$, which is the energy to free an electron from the CB edge. Using

$$E_\mathrm{C} - E_\mathrm{F} = kT \cdot \ln\left(\frac{N_\mathrm{C}}{n}\right) \tag{1.30}$$

$$E_\mathrm{F} - E_\mathrm{V} = kT \cdot \ln\left(\frac{N_\mathrm{V}}{p}\right) \tag{1.31}$$

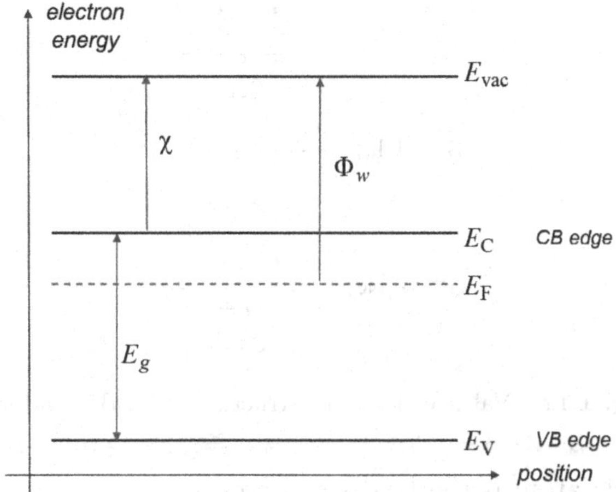

Fig. 1.11. Diagram of electron energy in a semiconductor.

it is clear that Φ_w depends on the carrier concentration.

$$\begin{aligned}\Phi_w &= E_{\text{vac}} - E_{\text{F}} = E_{\text{vac}} - E_{\text{C}} + (E_{\text{C}} - E_{\text{F}}) \\ &= \chi + kT \cdot \ln\left(\frac{N_{\text{C}}}{n}\right) = \chi - kT \cdot \ln\left(\frac{N_{\text{V}}}{p}\right) - E_g \qquad (1.32)\end{aligned}$$

1.2.4. *Impurity Doping*

Boron (B) and phosphorous (P) atoms form electron acceptors or donors, respectively, when replacing Si atoms in pure Si. These dopant atoms have one missing/extra valence electron. The dopant atom easily ionizes, by either accepting an electron (B) from the VB, or donating an electron (P) into the CB, leaving the ionized cores stationary in the lattice. The electronic structure of Si, B, and P are shown in Fig. 1.12.

The Bohr model for the hydrogen atom allows a simple estimation of the ionization energy of a dopant. The effective mass for the appropriate band is substituted for the electron rest mass. The effect of screening within the solid is incorporated into the Coulomb potential by replacing the permittivity of free space, ε_0, with the permittivity for the material, ε^*. The ionization energy of a

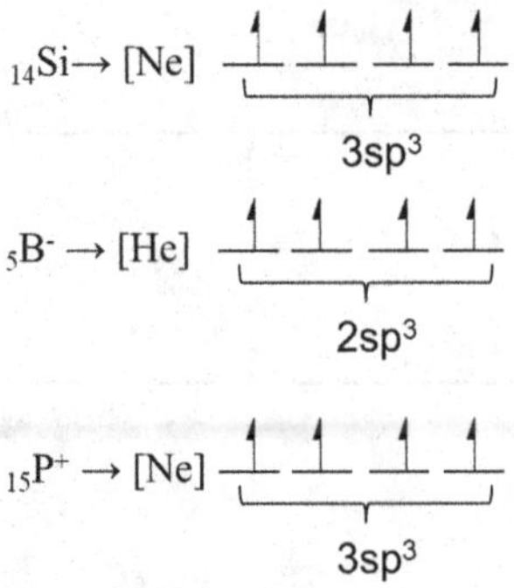

$_{14}Si \rightarrow [Ne]$ ↿ ↿ ↿ ↿ $3sp^3$

$_{5}B^{-} \rightarrow [He]$ ↿ ↿ ↿ ↿ $2sp^3$

$_{15}P^{+} \rightarrow [Ne]$ ↿ ↿ ↿ ↿ $3sp^3$

Fig. 1.12. Valence electronic structure of Si, B^-, and P^+.

hydrogen (H) atom is the Rydberg, given by

$$1\,\text{Ryd} = \frac{\hbar^2}{2m_0 \cdot a_0^2} = 13.6\,\text{eV} \tag{1.33}$$

The ionization energy of the dopant atom can then be written as:

$$V = \frac{\hbar^2}{2m^* \cdot (a_0^*)^2} = \left(\frac{m^*}{m_0}\right) \cdot \left(\frac{\varepsilon_0}{\varepsilon^*}\right)^2 \cdot (1\,\text{Ryd}) \tag{1.34}$$

This does not reflect any information on the chemical nature of the dopant atom, but simply provides an estimation of typical ionization energies (e.g., 45 meV for P and 45 meV for B in Si). Dopant levels indicated on the energy diagram convey the energies when the dopant atoms are neutral. The ionization energy of a dopant level corresponds to the separation of the level from the appropriate band edge. Shallow donor/acceptor levels are positioned slightly below/above the CB/VB.

For n-type material, the electron concentration corresponds to the concentration of ionized donors, i.e., $n = N_D^{(+)}$. For shallow donors, the Fermi level lies well below the donor level, so it is reasonable to assume that the donor level is fully ionized, giving $N_D^{(+)} = N_D$. We can then compute the Fermi level using

$$E_C - E_F = kT \cdot \ln\left(\frac{N_C}{n}\right) \tag{1.35}$$

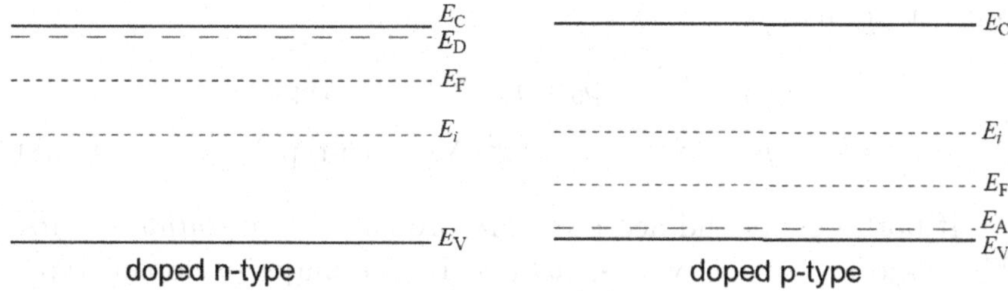

Fig. 1.13. Illustration of dopant levels donor (n-type) and acceptor (p-type) dopant levels within the bandgap of a semiconductor.

with $n = N_D$. The Fermi function evaluated at E_D gives the fraction of neutral donors. A more precision estimate of the ionized donor fraction is then

$$\frac{N_D^{(+)}}{N_D} = 1 - \frac{1}{1 + \exp(E_D - E_F)/KT} \tag{1.36}$$

We may then correct the estimated electron concentration using $n = N_D^{(+)}$. A self-consistent solution can be obtained by just a few iterations of this process.

Figure 1.13 shows some common doping situations found in silicon. The examples of Si doped with P and B are shown in Fig. 1.13. The ionization energy for P in Si is approximately $V_n \approx 0.045$ eV. The conduction band effective mass in Si is $m_C/m_0 = 1.1$. With $N_C = 2.9 \times 10^{19}\,\text{cm}^{-3}$, assuming complete ionization ($n = N_D$), we find $E_C - E_F = 0.188\,\text{eV}$. The fraction of ionized donors is $f(E_D - E_F) = 99.6\%$, or $n = 1.99 \times 10^{16}\,\text{cm}^{-3}$. Further iteration does not significantly alter these values, so the solution is self-consistent. Additional iterations may be needed for deeper donors and acceptors. The same procedure can be used for B-doping of silicon to calculate the free hole concentration.

If the medium is extrinsically doped with donors or acceptors, the minority-carriers are holes or electrons, respectively. As has been shown, the majority-carrier density produced by shallow donors or acceptors is approximately equal to the doping concentration. Therefore, the law of mass action gives the equilibrium free carrier concentrations:

$$n_0 = N_D; \quad p_0 = n_i^2/N_D, \quad \text{(n-type)} \tag{1.37}$$

$$p_0 = N_A; \quad n_0 = n_i^2/N_A, \quad \text{(p-type)} \tag{1.38}$$

If both donors and acceptors are present, *compensation* occurs. The majority carrier type is dictated by the majority dopant type, with some portion of the majority dopants serving to neutralize minority dopants. For example, if $N_D > N_A$, the material is n-type with $n_0 = N_D - N_A$. Likewise, if $N_A > N_D$, the material is p-type with $p_0 = N_A - N_D$.

As has been shown, the equilibrium density of electron-hole pairs is constant at a given temperature in a semiconductor. The law of mass action arises from thermodynamics and relates the product of electrons and holes to the intrinsic density, which is specific to a material. The intrinsic density is independent of the doping level, provided that the semiconductor is nondegenerate. With the definition $n_i = \sqrt{n \cdot p}$, the intrinsic density becomes slightly reduced at higher doping levels, into the degenerate regime, and is generally a function of dopant level. Band-gap shrinkage can also be expected in heavily doped semiconductors. These effects are discussed in detail by Landsberg Ref. 3. The GaAs intrinsic density, at degenerate doping levels, has been discussed in the literature Ref. 4.

Nonequilibrium

Excess carriers can be electrically or optically injected into the semiconductor, creating a nonequilibrium concentration, such that

$$n \cdot p > n_i^2 \tag{1.39}$$

Thermodynamics drives the system toward the equilibrium described by (1.25). With steady-state injection, the equilibrium is approached by various recombination mechanisms, which will be discussed in this chapter. In an intrinsic material, or under *high-injection* conditions, $n \approx p$, and the recombination process is called bimolecular decay.

Bimolecular recombination rates will be discussed in detail later in this work.

Assume that Δn and Δp are the excess electron and hole pairs concentrations, at some coordinate x and time t, produced by electrical or optical injection. We can write:

$$\Delta n(x,t) = n(x,t) - n_0 \tag{1.40}$$

$$\Delta p(x,t) = p(x,t) - p_0 \tag{1.41}$$

where n and p are the total, non-equilibrium carrier concentrations, and n_0 and p_0 are the equilibrium concentrations. The excitation may be either a steady state or a fast, pulsed source. The solutions to the excess carrier densities will be a central topic of forthcoming sections. The measurement of these excess-carrier densities are a central topic in this work.

When excess carriers are present, the system, in a steady state, is described as being in quasi-equilibrium. In particular, we cannot assume equilibrium between the carrier populations in the CB and VB, although we continue to assume equilibrium within each band. We can encapsulate these effects by identifying E_{F_n} and E_{F_p}, called electron and hole quasi-Fermi levels, and temperatures T_n and T_p, describing electron and hole temperatures. However, we usually assume the carrier population within each band has reached thermal equilibrium with the lattice at temperature T, such that $T_n \approx T_p = T$. The appropriate Fermi function to use within each band is then

$$f_n(E) = \frac{1}{\mathrm{e}^{(E-E_{\mathrm{F}_n})/kT} + 1} \tag{1.42}$$

$$f_p(E) = \frac{1}{\mathrm{e}^{(E-E_{\mathrm{F}_p})/kT} + 1} \tag{1.43}$$

The equilibrium analysis presented to this point can encapsulate time-independent, non-equilibrium phenomena: the so-called *steady state*. This quasi-equilibrium situation is always present under illumination, or with applied voltage, assuming sufficient time has elapsed for transient effects to diminish. The electron and hole carrier concentrations with each band are

$$n = n_i \cdot e^{(E_{Fn} - E_i)/kT} = N_C \cdot e^{-(E_C - E_{Fn})/kT} \tag{1.44}$$

$$p = n_i \cdot e^{(E_i - E_{Fp})/kT} = N_V \cdot e^{-(E_{Fp} - E_V)/kT} \tag{1.45}$$

For particular, non-equilibrium carrier concentrations, we can find the quasi-Fermi levels using:

$$\begin{aligned} E_{Fn} &= E_i + kT \cdot \ln\left(\frac{n + \Delta n}{n_i}\right), \quad \text{(electrons)} \\ E_{Fp} &= E_i - kT \cdot \ln\left(\frac{p + \Delta p}{n_i}\right), \quad \text{(holes)} \end{aligned} \tag{1.46}$$

Notice that the splitting of the quasi-Fermi levels is

$$\Delta E_F = kT \cdot \ln \frac{(n + \Delta n) \cdot (p + \Delta p)}{n_i^2} \tag{1.47}$$

A non-zero value of this splitting indicates that the CB and VB carrier populations are not in equilibrium with one another and is, essentially, a measure of the overall deviation from equilibrium. Within the space-charge region of a p/n junction, it corresponds to the external voltage across the junction. The $n \cdot p$ product now depends on the quasi-Fermi-level splitting, according to the law of mass action:

$$n \cdot p = n_i^2 \cdot e^{(E_{Fn} - E_{Fp})/kT} \tag{1.48}$$

Splitting of Quasi-Fermi Level

To demonstrate quasi-Fermi level splitting, we will use p-type silicon as an example and calculate the effect. In the figure, we take the material as doped p-type to a level of 1×10^{16} cm^{-3}. The calculation of the quasi-Fermi levels is then made as the injection level is increased from 1×10^5 to 1×10^{15} cm^{-3}.

One sees from the figure that E_{Fn} emerges from the equilibrium Fermi level at very low injection and passes through the intrinsic level at about 1×10^{10} cm^{-3}, the intrinsic density. The quasi-Fermi level concept is very useful for material and device analysis. The splitting

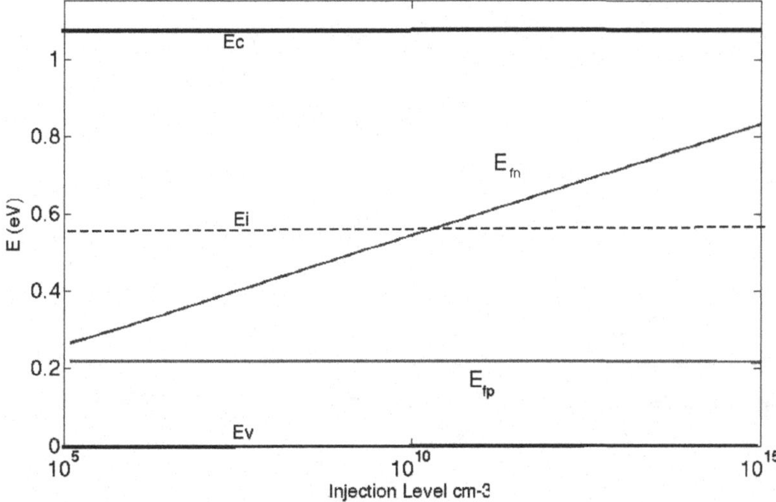

Fig. 1.14. A plot of the quasi-Fermi variation with injection level in p-type silicon doped to a level of $1 \times 10^{16}\,\text{cm}^{-3}$.

of the quasi-Fermi energy, ΔE_{F}, is related to the external voltage of solar cells.

$$qV_{oc} = \Delta E_{\text{F}}; \tag{1.49}$$

By measuring the steady state concentrations, Δn or Δp, at a known illumination intensity, one can predict the maximum voltage produced by a photovoltaic device based on that material using the quasi-Fermi level concept. This idea will be discussed in a later chapter.

References

1. Jacques I. Pankove, "Optical Processes in Semiconductors", Dover Publications, Inc., New York (1971).
2. S. Datta, Quantum Transport: Atom to Transistor (Cambridge University, Cambridge, UK, 2005).
3. P. Landsberg, Recombination in Semiconductors (Cambridge University, Cambridge, UK, 1991).
4. M. S. Lundstrom, Semiconductors and Semimetals 39 (Academic Press, Inc., Boston, 1993), pp. 194–255.

CHAPTER 2

Optical Absorption and Radiative Recombination

Figure 1.9 shows the E versus k diagram for a direct band gap semiconductor with a parabolic band structure. A transition from hole state in the valence band wave vector, k_p to an empty electron state in the conductor band wave vector, k_n, will be analyzed in terms of the density of states. We will define the zero of energy as the top of the valence band. The respective energies are:

$$E_i = -\frac{\hbar^2 k_p 2}{2m_p^*};$$

$$E_f = E_g + \frac{\hbar^2 k_n 2}{2m_n^*};$$

$$E_{h\nu} = E_f - E_i. \tag{2.1}$$

Here, we will assume that $T = 0$ such that the valence band is completely filled and the conduction band is completely empty. The optical transition requires conservation of total momentum. The momentum of the photon is relatively small and is negligible compared to the crystal momentum.

Therefore:

$$k_p = k_n = k; \tag{2.2}$$

The conservation of energy requires:

$$E_{h\nu} - E_g = \frac{\hbar^2 k^2}{2}\left(\frac{1}{m_n^*} + \frac{1}{m_p^*}\right) \equiv \frac{\hbar^2 k^2}{2}\frac{1}{m_a^*}. \tag{2.3}$$

Here, $1/m_a^*$ is the average inverse hole and electron mass.

The density of states in energy space and k-space are related as before:

$$N(E_{h\nu})dE_{h\nu} = \frac{8\pi k^2}{(2\pi)^3}dk. \tag{2.4}$$

Combining the above two equations, one can calculate the density of states for the absorption of a photon that connects E_i and E_f. The result is:

$$N(E_{h\nu}) = \frac{(2m_a^*)^{3/2}}{\pi^2\hbar^3}\sqrt{E_{h\nu} - E_g}. \tag{2.5}$$

The Beers law absorption coefficient is a product of the quantum mechanical matrix elements connecting the valence band and conduction band wave functions. The Beers law absorption coefficient can then be written as:

$$\alpha(E_{h\nu}) = C(m_n^*, m_p^*) \cdot N(E_{h\nu}) = \frac{(2m_a^*)^{3/2}}{\pi^2\hbar^3}\sqrt{E_{h\nu} - E_g}. \tag{2.6}$$

Here, C is a constant that contains the electron and hole effective masses and some fundamental constants.

Pankove approximated[1] the constant C with the result:

$$\alpha(E_{h\nu}) \cong \frac{q^2(2m_a^*)^{3/2}}{nch^2 m_n^*}\sqrt{E_{h\nu} - E_g} \tag{2.7}$$

Inserting the free electron masses as the effective masses and the index of refraction $n = 4$, one calculates:

$$\alpha(E_{h\nu}) \cong 2 \times 10^4\sqrt{E_{h\nu} - E_g}(\text{cm}^{-1}). \tag{2.8}$$

Here all energies are expressed in electron volts (eV). This value of absorption coefficient is relatively accurate for many direct band gap semiconductors. The absorption coefficient increases as the square root of the photon energy reflecting the increase of the density of states as k^2.

2.1. Radiative Recombination

Radiative recombination is the inverse process to optical absorption. An electron in the conduction band makes a transition to an empty valence band state (or recombines with a hole). This process produces a photon equal to that of the energy difference between the initial and final state. W. Van Roosbroeck and W. Shockley[2] proposed that in equilibrium, the blackbody rate of electron-hole generation must equal the rate of spontaneous radiative recombination. They performed a detailed balance calculation of these processes that resulted in the well-known van Roosebrocek-Shockley relationship. Their result is:

$$R(E_{hv})dE_{hv} = \frac{\alpha(E_{hv}) \cdot 8\pi E_{hv}^2 n^2}{h^3 c^2 \left[\exp\left(\frac{E_{hv}}{KT}\right) - 1\right]} dE_{hv} \tag{2.9}$$

Here, E_{hv} is the photon energy and $\alpha(E_{hv})$ is the Beers law absorption coefficient. Also, $R\ (E_{hv})$ is the luminescence intensity for that particular photon energy. We see that the emission intensity if proportional to the absorption coefficient. That mechanism underlies the much stronger light emission properties of direct band gap materials such as GaAs and related compounds.

The total emission rate, R, is obtained by integrating over the entire emission band.

$$R = \int_{E_g}^{\infty} \frac{\alpha(E_{hv}) \cdot 8\pi E_{hv}^2 n^2}{h^3 c^2 \left[\exp\left(\frac{E_{hv}}{KT}\right) - 1\right]} dE_{hv}. \tag{2.10}$$

Here, the lower limit is taken as E_g as $\alpha(E_{hv})$ is assumed to be zero at smaller energies. It is convenient to change to a dimensionless variable, $U = E_{hv}/KT$, for the integration. The total, equilibrium recombination rate, R_e, is:

$$R_e = \frac{8\pi n^2(KT)^3}{c^2h^3}\int_{u_g}^{\infty}\frac{\alpha(u)u^2du}{\exp(u)-1}. \tag{2.11}$$

Here, $u_g = E_g/KT$.

The nonequilibrium radiative recombination rate is then:

$$R = \frac{np}{n_i^2}R_e. \tag{2.12}$$

and,

$$B \equiv \frac{R_e}{n_i^2}. \tag{2.13}$$

B is defined as the radiative coefficient and is specific to a material. It decreases by three orders of magnitude from direct band gap to indirect band gap materials.

When excess carriers, Δn and Δp, are injected into a semiconductor, the luminescence intensity is:

$$I_{\text{PL}} = B(np - n_i^2) = B(n_0\Delta p + p_0\Delta n + \Delta p \cdot \Delta n);$$

where:

$$n_0p_0 = n_i^2. \tag{2.14}$$

Here, IL may be generated by optical, electrical or other methods of excess carrier generation. These results will be expanded later to describe transient and steady state PL measurements.

The blackbody or "dark" radiation is given by:

$$I_{bb} = Bn_i^2. \tag{2.15}$$

where B has units of cm^3s^{-1}. The values of B have been determined by both calculation and measurement by various workers. The most prevalent semiconductors in current technology are silicon and GaAs and related III-V compounds. The current accepted by values are given approximately by the values:

$$B_{Si} = 1 \times 10^{-15}\,\mathrm{cm^3\,s^{-1}};$$

$$B_{\mathrm{GaAs}} = 1.5 \times 10^{-10}\,\mathrm{cm^3\,s^{-1}}.$$

The calculated black body emission for each material is:

$$I_{bb}(\mathrm{Si}) = 1 \times 10^5 \text{ photons cm}^{-3}\,\mathrm{s}^{-1};$$

$$I_{bb}(\mathrm{GaAs}) = 1 \times 10^2 \text{ photons cm}^{-3}\,\mathrm{s}^{-1}.$$

Extra external injection increases the PL intensity by many orders of magnitude.

2.2. Calculation and Measurement of the Radiative Coefficient

It has been shown that the photon emission rate for band-to-band transitions is given by:

$$R_{\mathrm{L}} = Bnp(\mathrm{cm^3\,s^{-1}}) \tag{2.16}$$

Here p is the hole density, n is the electron density, and B is a coefficient that is specific to a given material. The B-coefficient may be calculated by summing the dipole matrix elements connecting the valence and conduction bands. B has the units of $\mathrm{cm^3/s}$ and determines the recombination rate of a specific material. B is many orders of magnitude larger for direct band gap than for indirect band gap semiconductors. A number of calculations have been made over the past five decades that have served to guide the experimental work. In recent years, a number of measurements have been made on the more popular materials that confirm the values for use in device design, as well as, comparison with theory.

2.3. Theory

Calculations by Dumke[3] provided the first calculations of B for both direct and indirect band gap semiconductors.

A calculation by Hall[4] provided the B-coefficient for direct semiconductors in terms of the band effective masses, band gap, dielectric constant, and temperature as:

$$B = 0.58 \times 10^{-12}\sqrt{\varepsilon}\left(\frac{1}{m_p + m_n}\right)^{1.5}$$

$$\times \left(1 + \frac{1}{m_p} + \frac{1}{m_n}\right)\left(\frac{300}{T}\right)^{1.5} E_g^2(\mathrm{cm}^3/\mathrm{s}) \qquad (2.17)$$

where ε is the dielectric constant, m_p and m_n are the hole and electron effective masses, respectively, in units of the free electron mass. Also T is the absolute temperature, and E_g is band gap in electron volts. Applying this equation to GaAs, and using effective masses $m_n = 0.08$ and $m_p = 0.5$, the calculation derives B as $1.4 \times 10^{-10}\,\mathrm{cm}^3/\mathrm{s}$.

Garbuzov[5] described a simple quantum mechanical calculation for direct band-gap semiconductors and obtained the following expression for the B-coefficient:

$$B = 3 \times 10^{-10}\left(\frac{E_g}{1.5}\right)^2\left(\frac{300}{T}\right)^{1.5} (\mathrm{cm}^3/\mathrm{s}) \qquad (2.18)$$

Putting in the parameters appropriate to GaAs at room temperature, one gets $B \sim 2.7 \times 10^{-10}\,\mathrm{cm}^3/\mathrm{s}$.

Quantum mechanical calculations of the B-coefficients were performed by Stern[6] and by Casey and Stern.[7] Thus, Stern concluded that B is not constant but varies weakly with carrier concentration.

Casey and Stern calculated the B-coefficient for p-GaAs over a range of doping levels. They included the doping concentration dependence of the energy gap and the energy-dependent matrix elements for the dipole transition. Stern[6] calculated the B-coefficient for p-type ($p = 4 \times 10^{17}\,\mathrm{cm}^{-3}$) GaAs at electron injection levels ranging from $n = 0$ to $n = 2.5 \times 10^{18}\,\mathrm{cm}^{-3}$. He also finds that $B \sim 2 \times 10^{-10}\,\mathrm{cm}^3/\mathrm{s}$ injected electron densities less than $1 \times 10^{18}\,\mathrm{cm}^{-3}$. Near $n = 2.5 \times 10^{18}\,\mathrm{cm}^{-3}$, the calculations show that B decreases to about $1 \times 10^{-10}\,\mathrm{cm}^3/\mathrm{s}$. In summary, their calculations indicated that B (p-GaAs) ranged from about $2 \times 10^{-10}\,\mathrm{cm}^3/\mathrm{s}$ ($\sim 1 \times 10^{18}\,\mathrm{cm}^{-3}$) to $1 \times 10^{-10}\,\mathrm{cm}^3/\mathrm{s}$ ($1 \times 10^{19}\,\mathrm{cm}^{-3}$).

H. van Cong[8] calculated the B-coefficient (77 Kelvin) for n-GaAs and predicted a minimum in the radiative lifetime at $n \sim 6\times10^{17}\,\text{cm}^{-3}$ which is the onset of degeneracy. These calculations predict that B decreases dramatically at degenerate doping levels because of the band-tailing effects.

Pankove[1] and Landsberg[9] provide the estimated B-coefficients for a number of common semiconductors in their classic books.

2.4. Experimental Results

Determination of the B-coefficient by measurement of carrier lifetime materials is complicated by the self-absorption of emitted photons and subsequent re-emission. This is an important effect in in direct band gap materials but appears to be inconsequential in indirect material because of the shift of emission peak to the long wavelength side of strong absorption. In the case of strong self-absorption, the measured lifetime becomes much larger than the radiative lifetime of 1/BN. This photon regeneration has been described as photon recycling in the literature. When photon recycling dominates, the correct radiative lifetime must be extracted from the data by other methods. As the effect increases with sample thickness, measuring a series of films with various thicknesses enables an extraction of the correct radiative lifetime.

The first systematic studies of the B-coefficient in GaAs were made by Nelson[10] and Nelson and Sobers.[11,12] They made systematic lifetime studies of p-type, isotype heterostructures as a function of carrier concentration. The DHs were grown by liquid phase epitaxy (LPE) had the structure p-$Al_{0.5}Ga_{0.5}As$/p-GaAs/p-$Al_{0.5}Ga_{0.5}As$. They used time-correlated single photon counting (to be described in a latter section) and calculated the photon recycling factor for the DH devices as a function of thickness (d). Nelson[10] systematically measured the PL lifetime as a function of thickness for p-type DHs with doping levels of 5×10^{15}, 2.9×10^{16}, and $1.7 \times 10^{17}\,\text{cm}^{-3}$. From the slope, they determined the B-coefficient.

The photon recycling factor is a function of structure and thickness. It is represented by the factor $\phi(d)$, where d is the thickness

Table 2.1. The B-coefficient of four semiconductors.

	B($cm^{-3}s$)	Reference
Silicon	2×10^{-15}	Pankove[1] (calculated)
Germanium	3.4×10^{-14}	Pankove[1] (calculated)
GaAs	2×10^{-10}	Nelson and Sobers[11]
	2.15×10^{-10}	Ahrenkiel[13] *et al.*
In(0.53)Ga(0.47)As	1.43×10^{-10}	Ahrenkiel[14] *et al.*

of the film. The effective radiative lifetime is then given by:

$$\tau_{\mathrm{PL}} = \frac{\phi(d)}{\mathrm{BN}}. \tag{2.19}$$

And the PL lifetime is larger than the radiative lifetime by the factor $\phi(d)$. Photon recycling will be described in a later section.

Some known values are shown in Table 2.1.

The calculated radiative lifetime is shown in curve B. The $\phi(d)$ coefficient is found by growing DHs of constant doping and variable thickness, d. The coefficient, ϕ, has also been calculated by numerous researchers as will be shown in a later section.

Lush and coworkers[15,16] performed extensive studies on hole lifetimes in p-MOCVD DHs over the concentration range of $1.3 \times 10^{17}\,\mathrm{cm}^{-3}$ to $3.8 \times 10^{18}\,\mathrm{cm}^{-3}$. For each doping concentration, five DH devices were grown with thicknesses ranging from $0.25\,\mu$m to $10\,\mu$m.

Figure 2.1 plots the PL lifetime of 0.25–0.50 μm and 8–10 μm DH structures from each doping concentration. The solid line is a plot of the radiative lifetime again assuming $B = 2 \times 10^{-10}\,\mathrm{cm}^3/\mathrm{s}$. All of the data points fall on or above the radiative lifetime. The data points labeled A are from devices either 0.25- or 0.50-μm-thick, and the points generally lie slightly higher than τ_{R}. The thick devices labeled B are either 8.0- or 10.0-μm-thick. The PL lifetime of the 10-μm device from the $1.3 \times 10^{17}\,\mathrm{cm}^{-3}$ series is 12.2 times larger than the radiative lifetime. As the doping density increases, $\tau_{\mathrm{PL}}/\tau_{\mathrm{R}}$ decreases. This decrease is a result of the drop in $\phi(d)$ with increased doping and will be discussed later. The A data points are representative of the radiative lifetime and agree quite well with the accepted B-value.

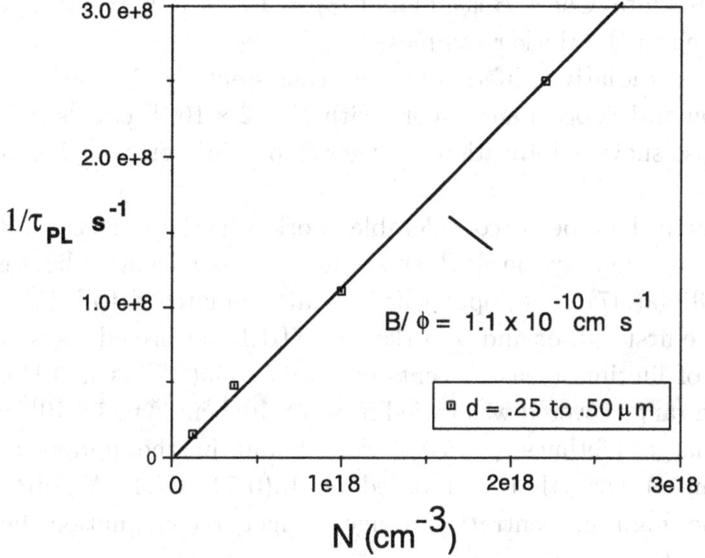

Fig. 2.1. The time-resolved PL lifetime of *p*-GaAs for a series of DHs of constant thickness, *d*, but doping ranging from about $2 \times 10^{15}\,\mathrm{cm}^{-3}$ to $1 \times 10^{19}\,\mathrm{cm}^{-3}$. The slope of $1/\tau_{\mathrm{PL}}$ versus N gives the *B*-coefficient for a set of films of constant thickness. The data shows the data of Ahrenkiel[13] and coworkers. The slope of data from the curve gives a B/ϕ value $1.1 \times 10^{-10}\,\mathrm{cm}^{-3}$. [Reprinted with permission from Ahrenkiel, R. K., Keyes, B. M., Lush, G. B., Melloch, M. R., Lundstrom, M. S., and MacMillan, H. F. (1992). *J. Vac. Sci. Technol.* **A10**, 990. Copyright [2019], American Vacuum Society.]

Detailed studies by Lush and all[17] show that there is a negligible component of bulk nonradiative recombination in this series of devices except at doping levels of $1 \times 10^{18}\,\mathrm{cm}^{-3}$ and higher. The higher doping levels showed an increase in lifetime at higher injection levels that is characteristic of SRH dominated recombination (see Fig. 2.1). This is similar to the case of *p*-type DH structures, in which the minority-carrier lifetimes in *n*-type epitaxial devices are controlled by radiative recombination. As verified by the data plotted in Fig. 2.1, the effective radiative lifetime may increase up to an order of magnitude due to photon recycling.

Figure 2.1. Curve A is a plot of predicted lifetime with $B = 2 \times 10^{-10}\,\mathrm{cm}^{-3}\mathrm{s}^{-1}$. As the samples are thin, photon recycling is minimal for these data. Curve B is a plot of $\tau_{PL} = 12.2 \times \tau_R$ because of photon recycling in the thicker samples.

The radiative lifetimes are consistent with earlier data (Nelson and Sobers) and theory with $B = 2 \times 10^{-10}\,\text{cm}^{-3}\text{s}^{-1}$. These data also show no indication of Auger recombination at higher doping levels.

There has been considerable work on the ternary InGaAs grown by epitaxy on InP substrates. In particular, the ternary In(0.53)Ga(.47)As is popular as it is lattice-matched to InP.

The first author and coworkers at NREL performed a systematic series of lifetime measurements on In(0.53)Ga(.47)As in 1998. The doping range covered was varied from $1 \times 10^{14}\,\text{cm}^{-3}$ to $1 \times 10^{19}\,\text{cm}^{-3}$. Very large lifetimes ($\sim 15\,\mu$s) were found in the undoped ($1 \times 10^{14}\,\text{cm}^{-3}$) DHs that had based on In(0.53)Ga(.47)As/InP DHs. At the high concentration range, Auger recombination became dominant.

The 2-μm-thick samples that were in the 10^{14} to 10^{17} doping range were measured by photoconductive decay (RCPCD). The lifetimes of the heavy doped samples were measured by the an up-conversion photon counting system at NREL. The composite data are shown in Fig. 2.2 below.

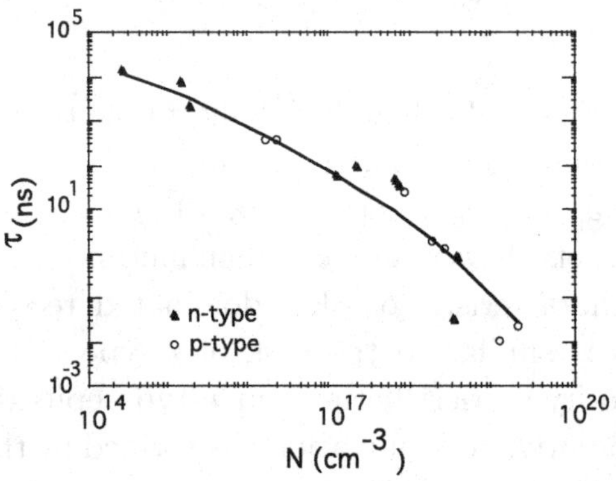

Fig. 2.2. The carrier lifetime for In(0.53)Ga(.47)As over a wide doping range.[14] [Reprinted from Lush, G. B., MacMillan, H. F., Keys, B. M., Ahrenkiel, R. K., Melloch, M. R., and Lundstrom, M. S. (1992). *Appl. Phys. Lett.* **61**, 2440, with permission of AIP Publishing.]

The data for the entire doping range was fit by the equation:

$$\frac{1}{\tau} = 2.11 \times 10^4 + 1.43 \times 10^{-10} N + 8.1 \times 10^{-29} N^2. \tag{2.20}$$

Clearly, the B-coefficient for In(0.53)Ga(.47)As is $1.42 \times 10^{-10}\,\text{cm}^{-3}\text{s}^{-1}$. This does not correct for photon recycling factors. If the latter correction made, B will be a factor of two larger, as the value here is really B/ϕ.

2.5. The Optical Generation Function

The Beers law absorption equation is given by:

$$I(x) = I_0 \exp(-\alpha x).$$

where α is the absorption coefficient of the material at a specific wavelength in units of cm^{-1}. Typical units for I_0 are photons $\text{cm}^{-2}\,\text{s}^{-1}$. On can easily calculate the number of e-h pairs (ΔI) in a volume of unit cross section and depth Δx.

$$\begin{aligned} \Delta I &= I_0(\exp(-\alpha x) - \exp(-\alpha(x + \Delta x)); \\ &= \alpha I_0 \exp(-\alpha x)\Delta x. \end{aligned} \tag{2.21}$$

The optical generation function is defined as:

$$G(x,t) = \alpha I_0 \exp(-\alpha x) f(t). \quad \text{photons/cm}^{-3}\,\text{s}^{-1}.$$

where $f(t)$ describes any time dependence of the excitation source. For pulsed excitation, one usually uses the impulse function as $f(t)$.

$$G(x,t) = \alpha I_0 \exp(-\alpha x)\delta(t). \tag{2.22}$$

This function will be used extensively for the forthcoming calculations.

References

1. Pankove, J. I. (1971). In *Optical Processes in Semiconductors*, Dover Publications, New York.
2. van Roosbroeck, W. and Shockley, W. (1954). *Phys. Rev.* **94**, 1558.
3. Dumke, W. P. (1957). *Phys. Rev.* **105**, 139.
4. Hall, R. N. (1960). *Proc. Inst. Elect. Eng.* **106B**, Suppl. 17, 983.

5. Garbuzov, D. Z. (1982). *J. Luminescence* **27**, 109.
6. Stern, F. (1976). *J. Appl. Phys.* **47**, 5382 .
7. Casey, H. C., Jr. and Stern, F. (1976). *J. Appl. Phys.* **47**, 631.
8. van Cong, H. J. (1981). *Phys. Chem. Solids* **42**, 95.
9. Landsberg, P. T. (1991). In *Recombination in Semiconductors*, Cambridge University Press, Cambridge.
10. Nelson, R. J. (1978). *J. Vac. Sci. Technol.* **15**, 1475.
11. Nelson, R. J. and Sobers, R. G. (1978). *J. Appl. Phys.* **49**, 6103.
12. Nelson, R. J. and Sobers, R. G. (1978). *Appl. Phys. Lett.* **32**, 761.
13. Ahrenkiel, R. K., Keyes, B. M., Lush, G. B., Melloch, M. R., Lundstrom, M. S., and MacMillan, H. F. (1992). *J. Vac. Sci. Technol.* **A10**, 990.
14. Ahrenkiel, R. K., Ellingson, R., Johnston S., and Wanlass, M. (1998). *Appl. Phys. Lett.* **72**, 3470.
15. Lundstrom, M. S. (1991). In *Proceedings of the Twenty-Second IEEE Photovoltaic Specialists Conference*, Las Vegas, Nevada, IEEE, New York, p. 182.
16. Lush, G. B. and Lundstrom, M. S. (1991). *Solar Cells* **30**, 337.
17. Lush, G. B., MacMillan, H. F., Keys, B. M., Ahrenkiel, R. K., Melloch, M. R., and Lundstrom, M. S. (1992). *Appl. Phys. Lett.* **61**, 2440.

CHAPTER 3

Mobility and Defect Recombination

During current flow, charge carriers, called quasi-particles, exhibit net motion through the solid. The electron and hole momentum in 1-D are simply:

$$p = hk = \frac{m_C}{h} \cdot \frac{d}{dk} E_C(k), \quad \text{(electrons)} \tag{3.1}$$

$$p = -hk = -\frac{m_V}{h} \cdot \frac{d}{dk} E_V(k), \quad \text{(holes)} \tag{3.2}$$

The corresponding speeds (velocities) are:

$$v = \frac{1}{h} \cdot \frac{d}{dk} E_C(k) = \frac{p}{m_C}, \quad \text{(electrons)} \tag{3.3}$$

$$v = -\frac{1}{h} \cdot \frac{d}{dk} E_V(k) = \frac{p}{m_V}, \quad \text{(holes)} \tag{3.4}$$

The Drude model is a semi-classical method that provides a simple description of conduction. In this model, we analyze the scattering of free quasi-particles (electrons and holes) from either lattice vibration (phonons) or point defects. Dislocations also act as scattering sites. Here we also assume that the scattering time is a constant and is not a function of other parameters such as particle velocity. In the model, a uniform electric field accelerates the carrier for a time, τ, and then scattering randomizes the motion. There is constant force on the free particle produced by an electric field.

$$F_C = -qE = m_C a_C \tag{3.5}$$

$$F_V = +qE = m_V a_V \tag{3.6}$$

We will assume a scattering rate $P = 1/\tau$. The number of unscattered free carriers at time t is:

$$n(t) = n_0 \exp(-t/\tau) \tag{3.7}$$

The product of the change at time t_i in the number of scattered particles Δn_i with their velocity v_i, in the time interval Δt is

$$\Delta n_i \cdot v_i = \frac{qEt_i}{m^*} \cdot n_0 \cdot \exp(-t_i/\tau) \cdot \Delta t. \tag{3.8}$$

(Here, m^* is the appropriate effective mass.) The average velocity of a free carrier, called the *drift* velocity, is then given by the ratio:

$$V_d = \frac{\frac{qEn_0}{m^*} \cdot \int_0^\infty t \exp(-t/\tau)dt}{n_0 \cdot \int_0^\infty \exp(-t/\tau)dt} = \frac{q\tau E}{m^*} \tag{3.9}$$

The mobility is defined as:

$$\mu_d = \frac{V_d}{E} = \frac{q\tau}{m^*} \tag{3.10}$$

The mobility has units of $\text{cm}^2/(\text{V} \cdot \text{s})$. A schematic representation of the random thermal velocity with a superimposed drift is shown in Fig. 3.1.

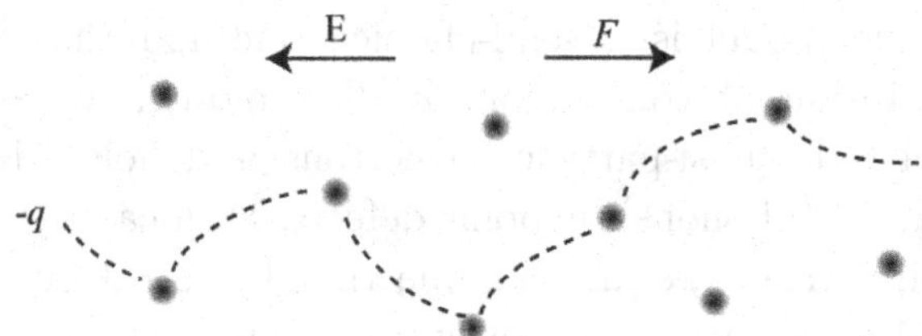

Fig. 3.1. Schematic of electron or hole motion with superimposed drift on random phonon scattering.

The mean current density is proportional to the drift velocity

$$\langle J \rangle = \pm qnV_d = \frac{q^2 nE}{m^*} \cdot \tau = \sigma E \tag{3.11}$$

The conductivity is:

$$\sigma = \frac{q^2 n\tau}{m^*} = q\mu n \tag{3.12}$$

With both carrier types present, the total current is:

$$J = J_n + J_p = q \cdot (\mu_n \cdot n + \mu_p \cdot p) \cdot E = \sigma E \tag{3.13}$$

and

$$\sigma = q \cdot (\mu_n \cdot n + \mu_p \cdot p)$$

3.1. Diffusion Equations

Electron and hole transport of charge also includes diffusion along concentration gradients. Fick's law that is used for diffusion of atoms defines these currents. The diffusion currents are given by:

$$J_n = qD_n \cdot \frac{dn}{dx}, \quad J_p = -qD_p \cdot \frac{dp}{dx} \tag{3.14}$$

A positive current is represent by a flow of positive charge in the $+x$ direction. The constants D_n and D_p are electron and hole diffusion coefficients, respectively. These are intimately related to the mobility because of the relationship to particle scattering by phonons or impurities.

3.2. The Einstein Relationship

The total current flow for electrons and holes is given by the current equations that combine drift and diffusion:

$$J_n = qn\mu_n E + qD_n \cdot \frac{dn}{dx} \tag{3.15}$$

$$J_p = qp\mu_p E - qD_p \cdot \frac{dp}{dx} \tag{3.16}$$

The E-field represents a change in electrical potential with position. When concentration gradients exist, there is corresponding *band bending* according to:

$$n(x) = N_C \cdot e^{-[E_C(x)-E_F]/kT} \tag{3.17}$$

Thus:

$$E = -\frac{1}{q}\frac{dE_C}{dx} \tag{3.18}$$

The band edges indicate the potential energies of electrons. Using Eq. (3.18), we see that the variation in $n(x)$ requires a corresponding variation in $E_C(x)$. The gradient in $n(x)$ is

$$\frac{dn}{dx} = -\left(\frac{1}{kT}\right)\cdot\frac{dE_C}{dx}\cdot n(x) = \frac{-qE}{kT}\cdot n(x) \tag{3.19}$$

In equilibrium, the total electron current is zero, and $J_n = 0$. Thus:

$$n\mu_n E - D_n \cdot \frac{dn}{dx} = 0 \tag{3.20}$$

Substituting, one gets the Einstein relationships for electrons and holes.

$$D_n = \frac{kT}{q}\mu_n \tag{3.21}$$

$$D_p = \frac{kT}{q}\mu_p \tag{3.22}$$

3.3. Scattering Mechanisms

The primary scattering mechanisms in single crystal materials are phonon and ionized impurity scattering. The latter may be composed of the intentional donor or acceptor additives. Unwanted ionized impurities act in the same manner and reduce the mobility. One common technique to assess the purity of laboratory materials is to cool the sample to very low temperatures and "freeze out" the phonon scattering, leaving the neutral and ionized impurities as the primary source for mobility reduction. In polycrystalline materials, the grain boundaries are a primary source of mobility reduction.

Finally, dislocations are also major sources of scattering in imperfect crystal. To evaluate the total mobility, one adds the scattering rates, which correspond to the inverses of the individual scattering lifetimes. Thus, the net scattering time can be evaluated as:

$$\frac{1}{\tau} = \frac{1}{\tau_{\mathrm{A}}} + \frac{1}{\tau_{op}} + \frac{1}{\tau_{\mathrm{D,A}}} + \frac{1}{\tau_I}. \tag{3.23}$$

Here τ is the total scattering time and τ_{A} and τ_{op} are acoustic and optical phonon events. The donor/acceptor and impurity scattering sources are included. The theoretical mobility of electrons and holes in silicon are calculated and shown in Fig. 3.2. Here, we assumed that the phonon scattering (300 K) and donor/acceptor scattering are the only sources. These data represent the maximum value that one can expect for silicon at a given doping level.

We see from the figure that donor/acceptor scattering starts to become a factor for doping levels greater than about $1 \times 10^{16}\,\mathrm{cm}^{-3}$.

3.4. The Continuity Equation

The continuity equation originally was derived from the theory of incompressible gas transport and describes the net flow of particles into a volume in terms of the rate of change of particle density.

$$\frac{d\rho}{dt} + \nabla \cdot \vec{J} = 0 \tag{3.24}$$

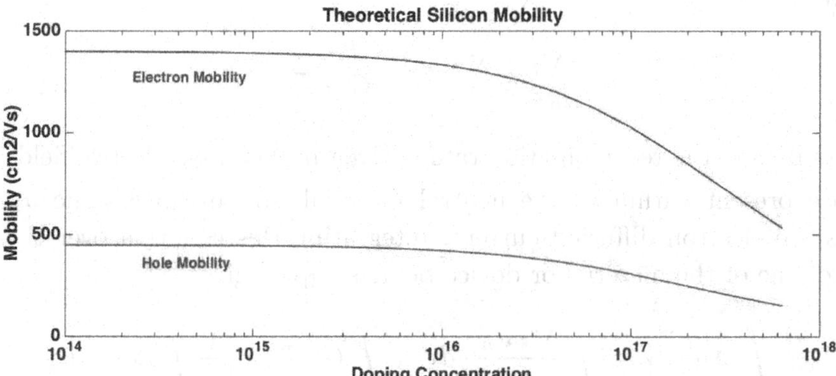

Fig. 3.2. Theoretical electron and hole mobility with only phonon and donor/acceptor scattering present.

Here, ρ is the particle density and $\vec{J}$ is the particle current. In this case,

$$\vec{J} = q\rho\vec{v} \tag{3.25}$$

where $\vec{v}$ is the particle velocity.

The continuity equation is modified for semiconductors to allow for generation and recombination of particles in the designated spatial volume. When the continuity equation is paired with the current equations, the result is

$$\frac{\partial n}{\partial t} = G_n - R_n + \frac{1}{q}\nabla \cdot \vec{J}_n \tag{3.26}$$

$$\frac{\partial p}{\partial t} = G_p - R_p - \frac{1}{q}\nabla \cdot \vec{J}_p \tag{3.27}$$

Here, n and p are the electron/hole densities, respectively, $\vec{J}_n$ and $\vec{J}_p$ are the current densities, G_n and G_p are the generation rates, and R_n and R_p are the recombination rates. For band-to-band absorption, we expect a single generation rate, so $G_n = G_p = G$. If this equation describes a specific spatial volume, the rate of change of particle density equals the divergence of the current, with the addition of generation and recombination. The recombination rate corresponds to the inverse lifetime (i.e., $R_n = 1/\tau_n$), which may be a function of the carrier density. For excess electrons in p-type material, this gives

$$\frac{\partial \Delta n}{\partial t} = \frac{\nabla \cdot \vec{J}_n}{q} + G - \frac{\Delta n}{\tau_n}. \tag{3.28}$$

As before, the recombination rate is given by $1/\tau_n$. No electric fields are present within charge neutral material, so we can assume $\vec{J}_n$ is the electron diffusion current. Integrating this equation over the volume of the material or device being sampled gives

$$\frac{\partial}{\partial t}\int_{\mathrm{V}} \Delta n \cdot dV = \int_{\mathrm{V}} \frac{\nabla \cdot \vec{J}_n}{q} \cdot dV + \int_{\mathrm{V}} G \cdot dV - \frac{1}{\tau_n}\int_{\mathrm{V}} \Delta n \cdot dV \tag{3.29}$$

This can be written as

$$\frac{\partial \Delta n_t}{\partial t} = \int_V \frac{\nabla \cdot \vec{J}_n}{q} \cdot dV + n_g - \frac{\Delta n_t}{\tau_n} \tag{3.30}$$

Here, Δn_t is the total number of excess electrons within the volume, and n_g is the number of photo generated electrons within the volume.

3.5. Recombination Mechanisms

The rate at which carrier concentrations are adjusted to their equilibrium values is characterized by a lifetime. We will identify the relevant recombination processes and identify the mechanisms that control lifetime. Three distinct bulk recombination mechanisms are identified in the following sections.

3.5.1. *Theory of the Shockley-Read-Hall Recombination Mechanism*

The Shockley-Read-Hall (SRH) model[1] has been used successfully for many decades to describe the role of a point defect in the nonradiative recombination process. The theory shows that a single energy level in the forbidden gap will promote the annihilation of electron-hole pairs, with the subsequent emission of long-wavelength phonons (represented as heat) in the material. The most efficient recombination is found to occur for defects that have an energy level near the middle of the forbidden gap. The SRH recombination process is derived by analyzing the equilibrium capture and emission rates for an electron and a hole. Using the schematic in Figure 3.3, the recombination center is characterized by five parameters:

E_{R}: The energy level of the center.
σ_n: The capture cross section for electrons by the center.
e_n: The emission rate for electrons from the center.
σ_p: The capture rate for holes by the center.
e_p: The emission rate for holes from the center.

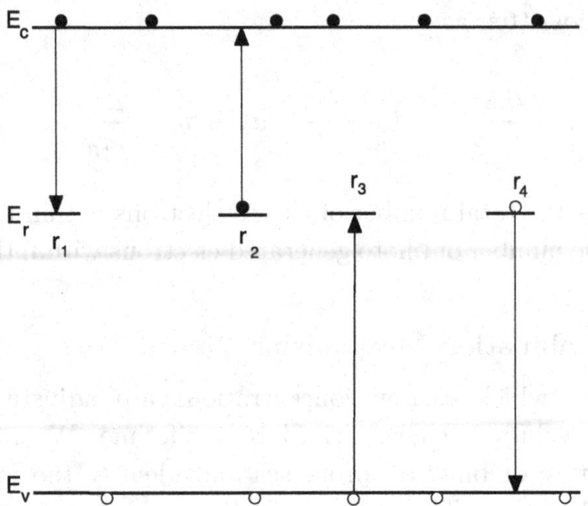

Fig. 3.3. Schematic representation of the Shockley Read Hall recombination process.

The electron and hole capture rates from single centers are related to the capture cross sections by

$$r_n = v_n \cdot \sigma_n \cdot n, \quad r_p = v_p \cdot \sigma_p \cdot p \tag{3.31}$$

Here v_n and v_p the thermal velocities of free electrons and holes, respectively.

Given a concentration N_R of recombination centers, with occupation probability $f(E_\mathrm{R})$, the volumetric rates for the four processes can be written as

$U_n = r_n \cdot N_\mathrm{R}^h$ Electron capture rate density.
$G_n = e_n \cdot N_\mathrm{R}^\mathrm{e}$ Electron emission rate density.
$U_p = r_p \cdot N_\mathrm{R}^\mathrm{e}$ Hole capture rate density.
$G_p = e_p \cdot N_\mathrm{R}^\mathrm{h}$ Hole emission rate density.

where $N_\mathrm{R}^\mathrm{e} = N_\mathrm{R} \cdot f_\mathrm{R}$, $N_\mathrm{R}^h = N_\mathrm{R} \cdot (1 - f_\mathrm{R})$ are the concentrations of occupied and unoccupied centers, respectively, using $f_\mathrm{R} = f(E_\mathrm{R})$. The rates of change of the free electron and hole concentrations, and the trapped carrier concentration, are

$$\frac{dn}{dt} = G_n - U_n \tag{3.32}$$

$$\frac{dp}{dt} = G_p - U_p \tag{3.33}$$

These rates vanish in equilibrium, indicating that $U_n = G_n$ and $U_p = G_p$. Using the equilibrium carrier concentrations n_0, p_0 to express the capture rates $r_n = v_n \cdot \sigma_n \cdot n_0$ and $r_p = v_p \cdot \sigma_p \cdot p_0$, the emission rates are found to be

$$e_n = r_n \cdot n_R/n_0,\ e_p = r_p \cdot p_R/p_0 \tag{3.34}$$

These refer, for convenience, to the free electron and hole concentrations that would occur in equilibrium if the Fermi level precisely coincided with the center energy (i.e., when the center is half full):

$$n_R = n_i \cdot \exp\left(\frac{E_R - E_i}{kT}\right) \tag{3.35}$$

$$p_R = n_i \cdot \exp\left(\frac{E_i - E_R}{kT}\right) \tag{3.36}$$

In non-equilibrium conditions, the electron and hole capture rates scale directly with the free-carrier concentrations, whereas the emission rates do not. Defining

$$\frac{1}{\tau_n} = v_n \cdot \sigma_n \cdot N_R \tag{3.37}$$

$$\frac{1}{\tau_p} = v_p \cdot \sigma_p \cdot N_R \tag{3.38}$$

the rates densities in non-equilibrium conditions become

$$\frac{dn}{dt} = \frac{n_R \cdot f_R - n \cdot (1 - f_R)}{\tau_n} \tag{3.39}$$

$$\frac{dp}{dt} = \frac{p_R \cdot (1 - f_R) - p \cdot f_R}{\tau_p} \tag{3.40}$$

$$\frac{d\rho_R}{dt} = -N_R \cdot \frac{df_R}{dt} \tag{3.41}$$

where $\rho_R = -N_R \cdot f_R$ is (within a constant offset) the concentration of charge carriers trapped in the center, which contribute to neither conductivity, nor band-to-band recombination.

All of the above rate densities depend on the occupation probability f_R of the center. Assuming the total charge density remains constant in a non-equilibrium scenario

$$\frac{dp}{dt} - \frac{dn}{dt} + \frac{d\rho_R}{dt} = 0 \tag{3.42}$$

The conventional SRH analysis assumes $d\rho_R/dt$ is much smaller than the rates associated with free electrons and holes. The SRH recombination rate is found by applying

$$U_{SRH} = U_n - G_n = U_p - G_p \tag{3.43}$$

Specifically

$$U_{SRH} = \frac{n \cdot (1 - f_R) - n_R \cdot f_R}{\tau_n} = \frac{p \cdot f_R - p_R \cdot (1 - f_R)}{\tau_p} \tag{3.44}$$

Given excess carrier concentrations $\Delta n = n - n_0$ and $\Delta p = p - p_0$, where n_0 and p_0 are the equilibrium concentrations, (3.44) is equivalent to

$$U_{SRH} = \frac{\Delta n}{\tau_n} \cdot (1 - f_R) = \frac{\Delta p}{\tau_p} \cdot f_R \tag{3.45}$$

We see that the minority-carrier lifetimes τ_n, τ_p are applicable in the limits of a deep level in doped material, i.e., $f_R \approx 0$ for n-type ($U_{SRH} = \Delta n/\tau_n$), or $f_R \approx 1$ for p-type ($U_{SRH} = \Delta p/\tau_p$), in which case the center generally contains a majority carrier. Thus, minority-carrier capture usually limits the recombination rate. The linear dependence on excess minority-carrier concentration corresponds to pure exponential decay of excess carrier densities. Hence, one usually refers to the observed inverse decay rate as the minority carrier lifetime. For high injection, the SRH recombination rate becomes non-linear with injection level, resulting in nonexponential decay. This effect will be the topic of a later section.

Consider the case of p-type material with an acceptor concentration of $N_a = 1 \times 10^{16}$ cm^{-3}, hole thermal velocity $v_p = 1 \times 10^7$ cm/s, and a center with hole capture cross section $\sigma_p = 1 \times 10^{-15}\,\text{cm}^{-2}$. The hole capture rate is then $r_p = 1 \times 10^8\,\text{s}^{-1}$, corresponding to a capture lifetime of 10 ns. The electron capture rate is smaller by a factor

$$\frac{r_n}{r_p} \approx \frac{\Delta n/\tau_n}{N_a/\tau_p} \tag{3.46}$$

Assuming $\tau_n \approx \tau_p$, electron capture limits the recombination rate density. This sequential process describes the essence of the SRH recombination mechanism.

The recombination "center" is usually an impurity atom, but it may be a molecular structure or an extended defect, like a dislocation. Substitutional gold (Au) in silicon is a well-known recombination center. The Au atom has exceptionally large capture cross sections ($\sigma_n \approx \sigma_p - 1 \times 10^{-14}\,\text{cm}^{-3}$) and the energy level is near the center of the forbidden gap. For that reason, Au has been used as a "lifetime killer" when the application is to make high-speed photodetectors.

Solving for the thermal distributions in (3.44) gives

$$f_\text{R} = \frac{\frac{p_\text{R}}{\tau_p} + \frac{n}{\tau_n}}{\frac{p+p_\text{R}}{\tau_p} + \frac{n+n_\text{R}}{\tau_n}}, \quad 1 - f_\text{R} = \frac{\frac{p}{\tau_p} + \frac{n_\text{R}}{\tau_n}}{\frac{p+p_\text{R}}{\tau_p} + \frac{n+n_\text{R}}{\tau_n}} \tag{3.47}$$

These are then combined to give the SRH recombination rate density as:

$$U_\text{SRH} = \frac{np - n_i^2}{(p + p_\text{R}) \cdot \tau_n + (n + n_\text{R}) \cdot \tau_p} \tag{3.48}$$

We can again separate the excess carrier densities Δn and Δp to give

$$U_\text{SRH} = \frac{(n_\text{O} + \Delta n) \cdot (p_\text{O} + \Delta p) - n_i^2}{(p_\text{O} + \Delta p + p_\text{R}) \cdot \tau_n + (n_\text{O} + \Delta n + n_\text{R}) \cdot \tau_p} \tag{3.49}$$

Consider low-level carrier injection in p-type material with an acceptor density of N_A. Assuming $\Delta p \ll N_\text{A}$, $\Delta n \gg (n_i/N_\text{A})^2 \cdot \Delta p$, and the presence of a deep level with $p_\text{R}, n_\text{R} \ll N_\text{A}$, the SRH rate becomes proportional the excess minority-carrier (electron) density, with no dependence on dopant concentration.

3.6. Theory of Surface and Interface Recombination

To this point, the SRH defects have been assumed to be uniformly distributed in the volume of the semiconductor. However, SRH defects are frequently localized at surfaces or interfaces between two regions of a material. The interface between the semiconductor and the outside environment is an inherent source of shallow and deep-level defects, origination from dangling bonds. These localized energy levels are generated at surfaces and interfaces because of the interruption of periodicity of crystalline materials and the resulting dangling bonds.

Early theory by Tamm[2] and by Shockley[3] showed that the interruption of periodicity of the crystal lattice produces a continuum of localized states in the forbidden gap. Tamm used the Kronig-Penney model of a one-dimensional crystal to show the quantum-mechanical origin of localized states in the forbidden gap. In the early literature, these states were called Tamm or Shockley states but today, they are called surface or interface states. The potential density of surface states is that of the two-dimensional density of surface atoms, about 10^{15} cm^{-2}. Early experimental work on silicon found that the crystal surface had a very high recombination rate, leading to the concept of surface recombination velocity (S or SRV).

As the surface acts as an infinite planar defect, it may also may act as a getter for impurities. These surface impurities also be a source of localized states and may coexist with dangling bond states. The interface in solid/solid systems results in bonding defects of greatly varying density. The terms surface and interface are often used somewhat interchangeably to describe the effects of the material boundary. The semiconductor interfaces greatly influence minority-carrier transport. Pioneering work by Bardeen[4] indicated that the potential barrier at semiconductor interfaces was more controlled by charges in interface states than by the contact potential difference of the materials. The most investigated interface in current semiconductor technology is that between silicon (Si) and thermally grown silicon dioxide (SiO_2). These states at the Si/SiO_2 interface are called surface or interface states and have been the subject of extensive experimental and theoretical studies. The reduction of

Si/SiO_2 interface states has been the focus of many years of research and development evolution of microelectronics. The density of surface states at the Si/SiO_2 interface has been reduced by many orders of magnitude by huge R&D investments into growth and processing techniques.

Early work indicated that the formation of SiO_2 reduced the interface state density by many orders of magnitude. The controlled growth[5] of SiO_2 on silicon, combined with various thermal annealing processes, reduces the Si/SiO_2 interface state density by many orders of magnitude. The removal of dangling bonds and their associated interface states is called passivation.

Silicon dioxide growth has been pivotal to the phenomenal development of the silicon technology and microelectronics. SiO_2 has a complementary application as a gate insulator in field-effect transistor (FET) technology. In metal-oxide semiconductor field-effect transistors (MOSFETs), the surface potential is intentionally modulated to control the current flow between source and drain. The SRV needs to be sufficiently small for source-drain current modulation to be effective. Therefore, the MOS device has primarily been limited to silicon-based technology because of the unique properties of the Si/SiO_2 interface.

Other semiconducting materials have been much more difficult to develop, largely because native oxide growth has generally not reduced surface or dangling bond states. Early experimental work[6] on the free surface of GaAs indicated that S exceeded 10^7 cm/s. In early transient decay experiments by Ehrhardt and coworkers[7] on GaAs wafers, an S of about 3×10^7 cm/s was measured. Early measurements indicated that surface effects dominated the volume minority-carrier lifetime in GaAs. The Illinois group of Zwicker and coworkers[8,9] was among the first to look at PL lifetimes in epitaxial GaAs. This group also attempted to reduce the dominant recombination velocity of the bare surface.

Early workers found that non-oxide passivation of the surface was the most productive method of SRV control. Dapkus and coworkers[10] reported one of the first confinement structure devices. They fabricated $n^+/n/n^+$ homostructures in which the n-layer is undoped GaAs. Confinement was produced by the upward band bending in the undoped layer, thereby driving electrons to the volume and repelling them from the surface. This device is limited in that the confinement barrier becomes ineffective for other than an undoped active layer.

A significant breakthrough in GaAs surface passivation came with the development of the heteroepitaxy and the resulting heterointerface. Pioneering work by the Illinois group of Keune *et al.*[11] used epitaxially grown $Al_xGa_{1-x}As$ as a wide bandgap confinement structure. Here, the conduction and valence band offsets provided the confinement of minority carriers to the active volume and repelled them from the surface. Because $Al_xGa_{1-x}As$ is closely lattice-matched to GaAs, dangling bonds are reduced or eliminated. In the early work, epitaxial $Al_xGa_{1-x}As$ became the standard confinement semiconductor material and allowed GaAs device technology to progress. The operation of early GaAs devices evolved because of the success of the epitaxial $Al_xGa_{1-x}As$/GaAs interface. Studies of the $Al_xGa_{1-x}As$/GaAs and newer epitaxial confinement systems will be described later. This technology is important for all devices that are dependent on minority-carrier transport. The surface or interface recombination velocity continues to be important in the operation of most compound semiconductor devices.

Surface recombination effects have been analyzed in detail in the book by Many, Goldstein, and Grover.[12] Although surface states are usually represented by a continuum of states in the forbidden gap, a model using a single level at energy E_R is useful in describing the recombination effects. In analyzing the surface or interface recombination effects, one can assume that these single-level SRH defects lie in a two-dimensional plane bounding the semiconductor. The limitations of the single-level approximation is compared with a more exact continuum model in a later paper by Landsberg and Brown.[13]

In Figure 3.4, we used the three dimensional model with SRH centers localized within a thin layer of width Δt at a 2-D interface.

The areal SRH recombination rate is found by multiplying the volume rate by the film thickness. We will then describe this thin film as a surface.

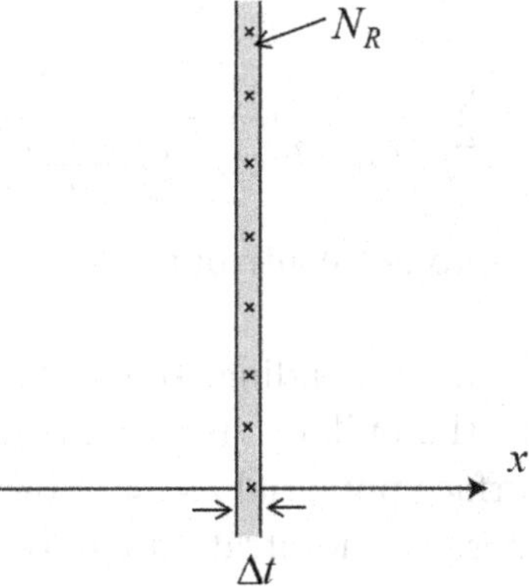

Fig. 3.4. Representation of the model used to calculate the surface recombination velocity using a finite volume containing point defects.

First consider the areal recombination rate within a layer of width Δt. For generality, assume an excess minority-carrier density of $\Delta\rho$ and a majority-carrier density N. For simplicity, assume the electron and hole capture lifetimes are equal, i.e., $1/\tau_n = 1/\tau_p = \sigma \cdot v_{th} \cdot N_{\mathrm{R}}$. Writing the interface, or surface, density of centers as $N_{\mathrm{S}} = N_{\mathrm{R}} \cdot \Delta t$, we obtain

$$U_{\mathrm{SRH}} \cdot \Delta t = \frac{\sigma \cdot v_{th} \cdot N_{\mathrm{S}} \cdot \Delta\rho}{1 + \frac{2n_i}{N} \cosh\left(\frac{E_{\mathrm{R}} - E_i}{kT}\right)} = S \cdot \Delta\rho \tag{3.50}$$

with units of $\mathrm{cm}^{-2}\mathrm{s}^{-1}$. The factor S is the surface recombination velocity:

$$S = \frac{\sigma \cdot v_{th} \cdot N_{\mathrm{S}}}{1 + \frac{2n_i}{N} \cosh\left(\frac{E_{\mathrm{R}} - E_i}{kT}\right)} \tag{3.51}$$

Notice that, for a midgap level (with $E_{\mathrm{R}} \approx E_i$), the recombination velocity becomes $S = \sigma_j \cdot v_{th} \cdot N_{\mathrm{S}}$. However, for most real surfaces or interfaces, S involves a summation over a number of near midgap states, which has been represented by a single state at energy E_{R}. Assuming that there are a collection of discrete energy levels, each with cross-section σ_j, areal density $N_{\mathrm{S},j}$, and energy $E_{\mathrm{R},j}$, Eq. (3.41) becomes:

$$S = \sum_j \frac{\sigma_j \cdot v_{th} \cdot N_{\mathrm{S},j}}{1 + \frac{2n_i}{N} \cosh\left(\frac{E_{\mathrm{R},j} - E_i}{kT}\right)} \tag{3.52}$$

Shallow states or traps do not contribute to S and may be neglected in the summation.

Because of surface band bending, the surface charge density ρ_S is not usually equal to the bulk or "flat band" density ρ. Defining a downward-bending surface potential as V_S,[14] band bending alters ρ_S with respect to the interior concentration ρ in the bulk, quasi-neutral region by

$$\rho_\mathrm{S} = \rho \cdot \exp\left(\frac{-qV_\mathrm{S}}{kT}\right) \tag{3.53}$$

The surface potential also shifts the intrinsic energy by qV_S, leading to a modified SRV of:

$$S = \frac{\sigma \cdot v_{th} \cdot N_s \cdot \exp\left(\frac{-qV_\mathrm{S}}{kT}\right)}{1 + \frac{2n_i}{N} \cosh\left(\frac{E_\mathrm{R} - E_i + qV_\mathrm{S}}{kT}\right)} \tag{3.54}$$

As before, if the energy level is located near midband at the surface ($E_\mathrm{R} \approx E_i - qV_\mathrm{S}$), Eq. (3.44) can be simplified:

$$S = \sigma \cdot v_{th} \cdot N_s \cdot \exp\left(\frac{-qV_\mathrm{S}}{kT}\right) \tag{3.55}$$

The utility of the SRV comes from applying continuity at an interface. For electrons in a p-type material, this is $\Delta J_n = q \cdot U_\mathrm{SRH} \cdot \Delta x$. If the interface is a free surface, passing from semiconductor to vacuum with increasing x, then $\Delta J_n = J_n|_+ - J_n|_- = 0 - J_n = -J_n$, which specifies a boundary value for electron current

$$J_n = -q \cdot S \cdot \Delta n \tag{3.56}$$

For holes in n-type material, the current at the boundary is

$$J_p = q \cdot S \cdot \Delta p \tag{3.57}$$

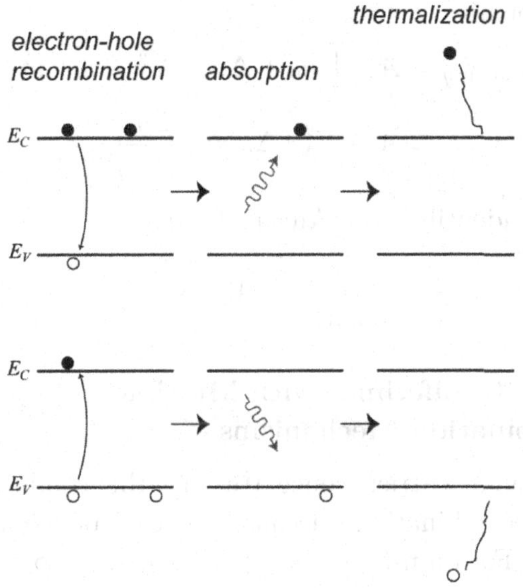

Fig. 3.5. Auger recombination processes for electrons and holes.

3.7. Auger Recombination

Auger processes involve the intraband exchanges of energy. Various processes are shown in Fig. 3.5. For example, a photon spontaneously emitted by an electron during recombination with hole can be absorbed by another electron, which is elevated above the CB edge, with the absorbed energy eventually being lost as heat during thermalization back to the band edge. We can anticipate the parametric dependence of the recombination rates:

$$U_{p,Aug} = A_p \cdot (n^2 \cdot p - n_0^2 \cdot p_0), \; U_{n,Aug} = A_n \cdot (n \cdot p^2 - n_0 \cdot p_0^2) \tag{3.58}$$

Due to the quadratic dependence on one carrier concentration, the disparity in Auger recombination rates between minority and majority carriers is even more exaggerated than for other mechanisms. Consider a p-type material with $p = N_\mathrm{A}$. The Auger recombination rate for electrons is:

$$\begin{aligned} U_{n,Aug} &= A_n \cdot \left[(n_0 + \Delta n) \cdot N_A^2 - n_0 \cdot p_0^2\right] \\ &\approx A_n \cdot N_A^2 \cdot \Delta n = \frac{\Delta n}{\tau_{n,Aug}} \end{aligned} \tag{3.59}$$

where we have identified the Auger lifetime:

$$\frac{1}{\tau_{n,Aug}} = A_n \cdot N_A^2 \tag{3.60}$$

3.8. Composite Lifetimes with Multiple Recombination Mechanisms

At any particular carrier concentration, the combined effects of the various recombination mechanisms can be combined into a single lifetime. For an n-type material, the excess hole concentration varies as:

$$\Delta p(t) = \Delta p(0) \cdot e^{-t/\tau_{p,\mathrm{SRH}}} \cdot e^{-t/\tau_{p,rad}} \cdot e^{-t/\tau_{p,Aug}} = \Delta p(0) \cdot e^{-t/\tau_p} \tag{3.61}$$

The composite lifetime is given by:

$$\begin{aligned} \frac{1}{\tau_p} = \frac{1}{\tau_{p,\mathrm{SRH}}} + \frac{1}{\tau_{p,rad}} + \frac{1}{\tau_{p,Aug}} &= \sigma_p \cdot v_p \cdot N_\mathrm{R} \\ &+ B_{rad} \cdot N_\mathrm{D} + A_{p,Aug} \cdot N_\mathrm{D}^2 \end{aligned} \tag{3.62}$$

Similarly, for electrons in a p-type material:

$$\begin{aligned} \frac{1}{\tau_n} = \frac{1}{\tau_{n,\mathrm{SRH}}} + \frac{1}{\tau_{n,rad}} + \frac{1}{\tau_{n,Aug}} &= \sigma_n \cdot v_n \cdot N_\mathrm{R} \\ &+ B_{rad} \cdot N_\mathrm{A} + A_{n,Aug} \cdot N_\mathrm{A}^2 \end{aligned}$$

In the context of the orders-of-magnitude variations in doping concentrations, we can summarize these mechanisms as follows:

1. intrinsic (no doping): SRH dominates
2. moderate doping: radiative dominates
3. high doping: Auger dominates

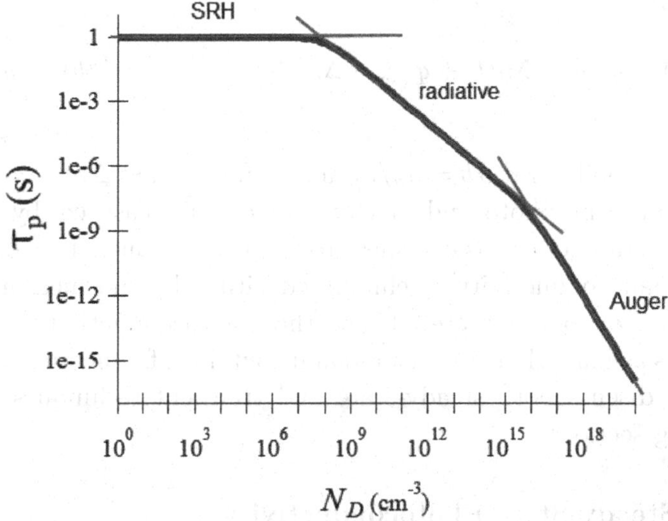

Fig. 3.6. Net hole lifetime for a hypothetical n-type semiconductor as a function of donor concentration.

Typical lifetimes over a large range of donor concentration are shown in Fig. 3.6.

3.9. Photoconductivity

External sources generated electron-hole pairs in a material or device, producing an additional component of conductivity. The most common laboratory light source is a monochromatic laser. The photoconductivity is then written with the photo-generated carriers, Δn and Δp.

$$\begin{aligned} \sigma_n &= q\mu_n \cdot (n + \Delta n) \\ \sigma_p &= q\mu_p \cdot (p + \Delta p) \end{aligned} \tag{3.63}$$

The photoconductivity is obtained by subtracting the "dark" conductivity from the total conductivity. The measurement of $\Delta\sigma$ is convoluted with any mobility changes as μ is a function of Δn. Therefore, mobility must usually be accounted for in the data analysis:

$$\Delta\sigma(t) = q \cdot \mu_n \cdot \Delta n(t) + q \cdot \mu_p \cdot \Delta p(t) = q\mu_n \cdot (1 + 1/b) \cdot \Delta n(t) \tag{3.64}$$

using the mobility ratio $b = \mu_n/\mu_p$ and assuming that $\Delta n(t) = \Delta p(t)$. The change in photoconductivity, $\Delta\sigma(t)$, is measured by either a steady-state or transient measurement technique. The dark or equilibrium conductivity is eliminated either by the measurement apparatus or by software. Then, the photoconductivity carriers produces a signal that is measured and contains lifetime information. We will discuss both steady-state and transient techniques in the following sections.

3.10. Steady-State Photoconductivity

Assuming photo generated excess carrier densities of $\Delta n = \Delta p = G\tau$, the conductivity change is

$$\Delta\sigma = q(\mu_n + \mu_p) \cdot G\tau \tag{3.65}$$

Therefore, the steady state lifetime is found by measuring $\Delta\sigma$ and G.

$$\tau = \frac{\Delta\sigma}{q(\mu_n + \mu_p) \cdot G} \tag{3.66}$$

Experimentally, one usually measures the total photoconductivity change $\Delta\sigma_{tot}$ of a wafer, while illuminating an area A. The wafer thickness, W, is measured and $\Delta\sigma$ is calculated by inserting the sample dimensions: i.e.

$$\Delta\sigma_t = \frac{\Delta\sigma^* W}{A}. \tag{3.67}$$

$$\tau = \frac{\Delta\sigma_t{}^* W}{qG(\mu_n + \mu_p)A}. \tag{3.68}$$

Here, $\Delta\sigma_t$ is the total sample photoconductivity measured by the apparatus. This analysis will be described in a later section in connection with the quasi-static photoconductive decay technique (QSSPC).[15]

3.11. High-Injection Effects on Mobility

High injection affects both the recombination lifetime and the mobility. The high injection effect on recombination rate will be discussed in a later section. Here we will look at the effects on mobility when the electron-hole concentration becomes sufficiently large that the Coulomb interaction between free carriers produces changes in the effective mobility. At low injection, the free carriers interact primarily with the background donors or acceptor and the external electric field supplied for the measurement. At higher injection levels, the Columbic attraction between the free electrons and holes competes with the applied electric field. A plasma forms and the free carriers move together because the internal field couples the carriers and opposes the external field. As the plasma concentration increases, the motion of the charge cloud slows in response to the electric field. This space-charge limited current (SCLC) mobility decreases with Δn and Δp. In this case, the effective mobility can decrease markedly as the internal Coulomb field couples the charge carriers. Following the fast-pulse generation of electron-hole pairs, one can write the continuity-current equation as:

$$\frac{\partial \Delta n}{\partial t} = \frac{1}{q}\nabla \cdot J_n - \frac{\Delta n}{\tau_n}, \quad \frac{\partial \Delta p}{\partial t} = -\frac{1}{q}\nabla \cdot J_p - \frac{\Delta p}{\tau_p} \tag{3.69}$$

One can write the one-dimensional form of the above for both electrons and holes;

$$\frac{\partial \Delta n}{\partial t} = D_n\frac{\partial^2 \Delta n}{\partial t^2} + \mu_n E\frac{\partial \Delta n}{\partial x} + \mu_n n\frac{\partial E}{\partial x} - \frac{\Delta n}{\tau_n} \tag{3.70}$$

$$\frac{\partial \Delta p}{\partial t} = D_p\frac{\partial^2 \Delta p}{\partial t^2} - \mu_p E\frac{\partial \Delta p}{\partial x} - \mu_p p\frac{\partial E}{\partial x} - \frac{\Delta p}{\tau_p} \tag{3.71}$$

We will apply the methods of van Roosebroeck[16] to analyze the effects of the internal electron-hole electric field. We insert the Poisson equation to replace the gradient of the electric field in both equations;

$$\frac{\partial E}{\partial x} = q\frac{(\Delta p - \Delta n)}{\varepsilon} \tag{3.72}$$

In radiative recombination, the electron and holes decay as pairs. We can then set $\Delta n = \Delta p$, and write

$$\frac{\partial \Delta n}{\partial t} = D_n \frac{\partial^2 \Delta n}{\partial t^2} + \mu_n E \frac{\partial \Delta n}{\partial x} - \frac{\Delta n}{\tau_n} \tag{3.73}$$

$$\frac{\partial \Delta p}{\partial t} = D_p \frac{\partial^2 \Delta p}{\partial t^2} - \mu_p E \frac{\partial \Delta p}{\partial x} - \frac{\Delta p}{\tau_p} \tag{3.74}$$

When we multiply the first equation by σ_p and the second by σ_n, we can add the two equations and the Poisson terms cancel to give

$$(\sigma_n + \sigma_n)\frac{\partial \Delta n}{\partial t} = -\Delta n \left(\frac{\sigma_n}{\tau_n} + \frac{\sigma_p}{\tau_p}\right) + (\sigma_n D_p + \sigma_p D_n)\frac{\partial^2 \Delta n}{\partial x^2} + (\sigma_p \mu_n - \sigma_n \mu_p) E \frac{\partial \Delta n}{\partial x} \tag{3.75}$$

Dividing by $\sigma_n + \sigma_p$, one can write the coupled-pair transport:

$$\frac{\partial \Delta n}{\partial t} = -\left(\frac{\frac{\sigma_n}{\tau_n} + \frac{\sigma_p}{\tau_p}}{\sigma_n + \sigma_p}\right) \cdot \Delta n + \left(\frac{\sigma_n D_p + \sigma_p D_n}{\sigma_n + \sigma_p}\right) \cdot \frac{\partial^2 \Delta n}{\partial x^2} + \left(\frac{\sigma_p \mu_n - \sigma_n \mu_p}{\sigma_n + \sigma_p}\right) \cdot E \cdot \frac{\partial \Delta n}{\partial x} \tag{3.76}$$

The coupled-pair transport parameter are written by comparison with the single particle transport:

$$\frac{1}{\tau_{\mathrm{A}}} = \frac{\frac{\sigma_n}{\tau_n} + \frac{\sigma_p}{\tau_p}}{\sigma_n + \sigma_p} \tag{3.77}$$

$$D_{\mathrm{A}} = \frac{\sigma_n D_p + \sigma_p D_n}{\sigma_n + \sigma_p} = \frac{n + p}{\frac{n}{D_n} + \frac{p}{D_p}} \tag{3.78}$$

$$\mu_{\mathrm{A}} = \frac{\sigma_p \mu_n - \sigma_n \mu_p}{\sigma_n + \sigma_p} = \frac{n - p}{\frac{n}{\mu_n} + \frac{p}{\mu_p}} \tag{3.79}$$

These are the ambipolar lifetime, diffusivity, and mobility.

Here, we will assume that the ionized donor/acceptor scattering is independent of doping type for each charge carrier, i.e.;

$$\mu_n = \mu_n(N), \ \mu_p = \mu_p(N) \tag{3.80}$$

where $N = N_{\rm A}, N_{\rm D}$. The electron-hole pairs move collectively due to the Coulomb attraction when the injected density becomes large compared to background doping. The response to the external field is that this ensemble moves more slowly and therefore the space-charge limited current slows relative to that of an individual particle.

Taking the case of a p-type silicon sample, one can calculate the ambipolar mobility over a range of doping and injection level. Using the parameters and looking at low injection, we know $p_0 = N_{\rm A}$ and $\Delta p = \Delta n \ll N_{\rm A}$.

$$\mu_{\rm A} \simeq \frac{-N_{\rm A}}{\frac{\Delta n}{\mu_p} + \frac{N_{\rm A}}{\mu_n}} \simeq -\mu_n. \tag{3.81}$$

At very high injection, $\Delta n = \Delta p \gg N_{\rm A}$. Then

$$\mu_{\rm A} \simeq \frac{-N_{\rm A}}{\frac{\Delta n}{\mu_p} + \frac{\Delta p}{\mu_n}} \simeq \frac{-N_{\rm A}}{\Delta n} \frac{\mu_n \mu_p}{\mu_n + \mu_p} \tag{3.82}$$

Assuming that $\mu_n \gg \mu_p$,

$$\mu_{\rm A} \simeq -\frac{N_{\rm A}}{\Delta p} \mu_p \tag{3.83}$$

In this case, the ambipolar mobility becomes much smaller the smallest individual mobility.

In the Fig. 3.7, curve A is a calculation of the ambipolar mobility of p-type silicon that is very lightly doped ($1 \times 10^{14}\,\rm cm^{-3}$) silicon. Curve B is a calculation for the case of moderate p-doping ($1 \times 10^{15}\,\rm cm^{-3}$) and curve C is moderately p-doped ($1 \times 10^{16}\,\rm cm^{-3}$). The space charge effects decrease as the background doping increases. The injection level is systematically increased from 2×10^{11} to $1 \times 10^{15}\,\rm cm^{-3}$. The largest value is that easily generated by laboratory laser sources. We see from the data that the mobility reduction is dramatic for high resistivity silicon whereas the mobility reduction

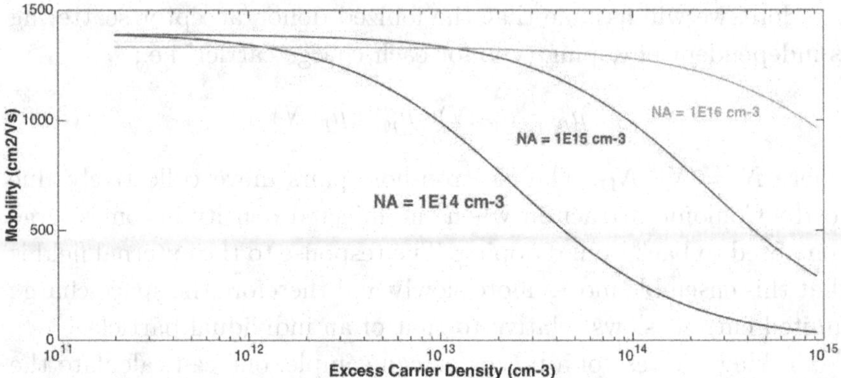

Fig. 3.7. Theoretical ambipolar mobility in silicon for electron-hole pairs at a range of background doping levels.

is only apparent at the $1 \times 10^{15}\,\text{cm}^{-3}$ level in the moderately doped case.

These mobility effects must be taken into account for the accurate interpretation of photoconductive data. Space charge effects are especially severe for the high resistivity samples. Examples of this effect will be shown in a later section of this work.

3.12. Summary

The mobility of carriers is controlled by a combination of phonon and impurity scattering. The latter includes mechanical defects such as dislocations, grain boundaries, and surfaces. At high injection, space effects that arise from the electrostatic coupling of electrons and holes and further reduce the mobility. This internal electric field counteracts the externally applied field to reduce the motion of free carriers. The space charge effects will become significant for devices that work under high-injection conditions.

References

1. Shockley, W. and Read, W. T., Jr. (1952). *Phys. Rev.* **87**, 835.
2. Tamm, I. E. (1932). *Z. Phys.* **76**, 849.
3. Shockley, W. (1932). *Phys. Rev.* **56**, 317.
4. Bardeen, J. (1947). *Phys. Rev.* **71**, 727.

5. Nicollian, E. H. and Brews, J. R. (1982). In *MOS (Metal Oxide Semiconductor) Physics and Technology*, John Wiley & Sons, New York, p. 645.
6. Casey, H. C., Jr. and Buehler, E. (1977). *Appl. Phys. Lett.* **30**, 2550.
7. Ehrhardt, A., Wettling, W., and Bett, A. (1991). *Appl. Phys.* **A53**, 123.
8. Zwicker, H. R., Keune, D. L, Holonyak, N., Jr., and Burnham, R. D. (1971). *Solid St. Electron* **14**, 1023.
9. Zwicker, H. R., Scifres, D. R., Holonyak, N., Jr., Dupuis, R. D., and Burnham, R. D. (1971). *Solid State Comm.* **9**, 587.
10. Dapkus, P. D., Holonyak, N., Jr., Burnham, R. D., and Keune, D. L. (1970). *Appl. Phys. Lett.* **16**, 93.
11. Keune, D. L., Holonyak, N., Jr., Burnham, R. D., Scifres, D. R., Zwicker, H. W., Burd, J. W., Craford, M. G., Dickus, D. L., and Fox, M. J. J. (1971). *J. Appl. Phys.* **42**, 2048.
12. Many, A., Goldstein, Y., and Grover, N. B. (1971). In *Semiconductor Surfaces*, North-Holland Publishing Company, Amsterdam, p. 194.
13. Landsberg, P. T. and Browne, D. C. (1988). *Semicond. Sci. Technol.* **3**, 193.
14. Dhariwal, S. R. and Mehrotra, D. R. (1988). *Solid-St. Electron.* **31**, 1355.
15. Sinton, R. A. and Cuevas, A. (1996). *Appl. Phys. Letters* **69**, 2510.
16. van Roosebroeck, W. (1953). *Phys. Rev.* **9**, 282.

CHAPTER 4

Carrier Diffusion and Confinement — Transient and Steady State Theory

4.1. Introduction

The measurement of carrier lifetime is a very important component of many of the current semiconductor technologies. This measurement is especially important in the exploding photovoltaic industry, where carrier lifetime is a key parameter in the compromise between material cost and device performance. Measuring carrier lifetime along the chain of device processing steps is a very desirable goal of many of the commercial photovoltaic companies. Early identification of the physical phenomena, that degrades the final device performance, is a very important cost-reduction activity. Production line compatibility requires fast measurement (i.e. throughput time) without becoming a burdensome retardant of production rate.

The rapid measurement of lifetime, and corresponding material quality, is probably best achieved by the Quasi-Steady state photoconductivity technique (QSSPC) for silicon materials. QSSPC has become an industry standard for the silicon photovoltaic industry. However, it has only been developed for silicon technology at this point in time.

The other application of lifetime techniques is the research laboratory where new materials are being invented/developed. In this case, determination of the dominant recombination mechanism(s) is the primary goal, as well as, the quality of new materials. In this application, acquisition time is much less consequential. On the other hand, the accuracy and depth of the analysis is much more important. Here, the transient techniques have an advantage as the time dependence of the carrier decay contains a lot of information about the nature of the recombination process. A great deal of emphasis will be placed on the decay function. In the experience of the first author, misinterpretation is a frequent event for transient measurements.

A variety of transient measurement techniques have been developed over the last few decades. Two of these have been commercialized; microwave reflection and time-resolved photoluminescence. These techniques will be discussed in great detail here.

When excess carriers are generated, a number of physical mechanisms exist that drive the system back toward equilibrium. These recombination mechanisms are the forces that result in the excess carriers having a finite lifetime.

We define a general recombination rate $R(n,p,t)$ as:

$$\frac{dn}{dt} = \frac{dp}{dt} = G(n,p,t) - R(n,p,t),$$

where;

$$R(n,p,t) = r(n,p) * (np - n_i^2) \qquad (\mathrm{cm}^{-3}\,\mathrm{s}^{-1}) \qquad (4.1)$$

Here, $r(n,p)$ is a function that represents the specific recombination mechanism with dimensions of $\mathrm{cm}^3\mathrm{s}^{-1}$. The function $G(n,p,t)$ represents the generation rate of electron-hole pairs by external sources such as absorbed photons. $G(n,p,t)$ is generally either a constant in time (the steady state case) or a nearly instantaneous pulse (the transient case).

Specific recombination mechanisms with include:

- Radiative recombination.
- Impurity or Shockley-Read-Hall (SRH) recombination.

- Auger recombination.
- Surface recombination.

When excess carriers are generated by some external steady-state source, $G(r)$, the equilibrium condition is described as:

$$\frac{dn}{dt} = \frac{dp}{dt} = 0,$$
$$R(n,p,t) = G(r,t). \tag{4.2}$$

Here $R(n,p,t)$ is the recombination rate of electron-hole pairs and equals the loss rate of the electron-hole pairs from the volume. In general, r is a function of n and p, but for radiative recombination r is a constant and specific to a given material. In the latter case, r is called the B or radiative coefficient. The value of the radiative (B-coefficient) for a given material will be analyzed in a later section.

The radiative case (r being a constant and independent of the carrier concentration) is particularly simple and will be discussed first. The functional form of r for the other common (4.3) recombination mechanisms will be discussed later in this section.

For the injection of Δn, Δp electron-hole pairs into p-type material, the recombination rate of minority electrons is given by:

$$\frac{dn}{dt} = \frac{d\Delta n}{dt} = G(r,t) - r\left[(N_{\mathrm{A}} + \Delta p)(n_0 + \Delta n) - n_i^2\right] = 0. \tag{4.3}$$

The term $\Delta p\, n_0$ is dropped because n_0 is usually very small compared to both N_{A} and Δp. In low-injection, the Δn^2 term can be neglected. For the steady state case, $d\Delta n/dt = 0$. Therefore:

$$rN_{\mathrm{A}}\Delta n = G,$$
$$\Delta n = \frac{G}{rN_{\mathrm{A}}} \equiv G\tau \tag{4.4}$$

Thus, τ is the recombination lifetime and τ is inversely proportional to $\mathrm{RN_A}$ in this case.

4.2. Transient Measurements

For the transient case, in which G is a short pulse and can be described by impulse function, $G_0\delta(t)$: i.e.

$$\begin{aligned} G &= G_0; \quad t = 0, \\ G &= 0; \quad t > 0. \end{aligned} \tag{4.5}$$

For $t > 0$,

$$\frac{d\Delta n}{dt} = -rN_{\mathrm{A}}\Delta n \tag{4.6}$$

The excess minority-carrier density then decays with a time constant τ. The decay time τ is the minority-carrier lifetime.

$$\Delta n(t) = \Delta n_0 \exp(-rN_{\mathrm{A}}t) = \Delta n_0 \exp(-t/\tau) \tag{4.7}$$

where again $\tau = 1/rN_{\mathrm{A}}$.

In the case of radiative recombination, r equals B and decreases by five orders of magnitude from direct to indirect bandgap semiconductors. Thus, GaAs, has strong light emission and is widely used for light emitting devices and diode lasers. Silicon has a B-coefficient about 1×10^{-5} that of GaAs. Thus, silicon is a weak light emitter, but has a long lifetime making it more suitable for microelectronics.

For the deep level defect recombination, the low-injection recombination rate $r(n, p)$ will be shown to be approximately $\frac{N_t \sigma_n v_{\mathrm{th}}}{N_{\mathrm{A}}}$ where N_t is the recombination center density. Also, σ_n is the capture cross-section for electrons, and v_{th} is the thermal velocity of the electron. In this case, the total recombination rate $R(n, p)$ is $N_t \sigma_p v_{\mathrm{th}} \Delta p$ and is independent of the doping level.

Rearranging Eq. (4.6), the lifetime is given by:

$$\tau = -\frac{\Delta n(t)}{\left(\frac{d\Delta n(t)}{dt}\right)} \tag{4.8}$$

When $\Delta n(t)$ data are acquired, the lifetime is obtained by plotting log $[\Delta n(t)]$ versus time and measuring the slope of the plot.

The quantity $r(n, p)$ will be discussed for radiative and nonradiative recombination mechanisms and may be a simple constant or a more complex function of material parameters. The concept of a single minority-carrier lifetime obviously applies only to low injection conditions for extrinsic or doped semiconductors. For certain situations such as high-injection, the recombination rate becomes nonlinear.

The surface or interface of a material is inherently defective and is described as a surface (or interface) recombination velocity or SRV. The symbol S will be used for surface recombination velocity here and has units of cm/s. The current into the surface defect states will be shown to be:

$$J_s = \Delta n^* S. \quad (\mathrm{cm}^{-2}\,\mathrm{s}^{-1}). \tag{4.9}$$

This component of recombination is routinely used in device analysis.

Applications of these results will underlie the remainder of this section.

4.3. The Current Equation

The electron and hole current densities are given by the current equations:

$$\begin{aligned}\bar{J}_n &= q\mu_n n\bar{E} + qD_n\nabla n\\ \bar{J}_p &= q\mu_p p\bar{E} - qD_p\nabla p\\ \bar{J}_t &= \bar{J}_n + \bar{J}_p\end{aligned} \tag{4.10}$$

where D_n, D_p are the electron and hole diffusion coefficients, respectively.

The diffusion coefficients are related to the respective mobility by the Einstein relationship:

$$\begin{aligned}D_n &= \frac{KT}{q}\mu_n\\ D_p &= \frac{KT}{q}\mu_p.\end{aligned} \tag{4.11}$$

Also $\nabla p, \nabla n$ are the gradients in the hole and electron concentrations, respectively.

The conductivity is defined as the current density per unit of applied electric field.

$$J \equiv \sigma E = (qn\mu_n + qp\mu_p)E. \tag{4.12}$$

Therefore: $\sigma = qn\mu_n + qp\mu_p \equiv \sigma_n + \sigma_p$.

4.4. Diffusion Length

To define diffusion length, assume that an injecting electron contact is placed at the surface of a p-type wafer as shown in Fig. 4.1. The contact maintains a constant density of electrons at the surface by a weak forward bias. Thus, the continuity equation becomes with $\rho = \Delta n$.

$$\rho(0) = \rho_0;$$

Assuming that the sample is infinitely thick, the solution to the continuity equation becomes:

$$\rho(x) = \rho_0 \exp(-x/\sqrt{D_n \tau}). \tag{4.13}$$

This behavior is shown schematically in Fig. 4.1 with the excess electron concentration decaying exponentially in a diffusion length, L_n.

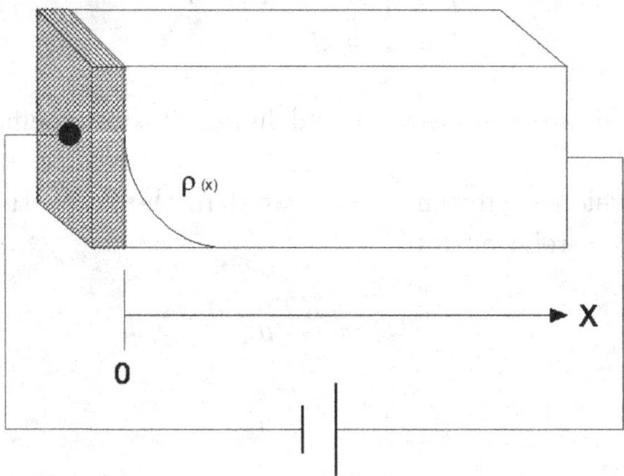

Fig. 4.1. Model used to calculate the diffusion length after injection of electrons in a p-type material.

For n-type and p-type semiconductors, the excess carrier density decreases with the electron or hole diffusion length, respectively.

$$L_n \equiv \sqrt{D_n \tau} \quad \text{and} \quad L_p \equiv \sqrt{D_p \tau}. \tag{4.14}$$

Here L_n and L_p define the electron and hole diffusion length for excess carrier density.

4.5. Confinement Structures

To analyze confinement structures, we combine the continuity equations for electrons and holes with the current equations and get:

$$\begin{aligned} \frac{\partial n}{\partial t} &= G_n - R_n + \nabla \cdot (\mu_n n \bar{E} + D_n \nabla n) \\ \frac{\partial p}{\partial t} &= G_p - R_p + \nabla \cdot (\mu_p p \bar{E} - D_p \nabla p). \end{aligned} \tag{4.15}$$

As the background carrier densities are constant in time, we write the above results in terms of excess carrier densities as functions of time:

$$\begin{aligned} \frac{\partial \Delta n}{\partial t} &= G_n - R_n + \nabla \cdot (\mu_n n \bar{E} + D_n \nabla n) \\ \frac{\partial \Delta p}{\partial t} &= G_p - R_p + \nabla \cdot (\mu_p p \bar{E} - D_p \nabla p). \end{aligned} \tag{4.16}$$

If electric fields are absent and the background doping is spatially constant, the one-dimensional combined equation for excess carriers is written as:

$$\begin{aligned} \frac{\partial \Delta n}{\partial t} &= D_n \frac{\partial^2 \Delta n}{\partial t^2} + G_n - \frac{\Delta n}{\tau_n} \\ \frac{\partial \Delta p}{\partial t} &= D_n \frac{\partial^2 \Delta p}{\partial t^2} + G_p - \frac{\Delta n}{\tau_p}. \end{aligned} \tag{4.17}$$

This equation is often called the diffusion equation. Here, Δn and Δp are the excess electron and hole densities. G is the generation and $1/\tau$ represents the electron and hole lifetimes. When this equation is applied to a specific spatial volume, the rate of change of particle density equals the divergence of the current, with the addition of generation and recombination. We represent the recombination rate by an inverse lifetime $(1/\tau_n)$ that may be a function of the carrier density. For excess electrons in p-type material, one can write the three dimensional equation in the following form:

$$\frac{\partial \Delta n}{\partial t} = \frac{\nabla \cdot \bar{J}_n}{q} + G - \frac{\Delta n}{\tau_n} \tag{4.18}$$

As before, the recombination rate is represented by, $1/\tau_n$. Also, J_n is the diffusion current and is the flow of electrons driven by the concentration gradient. When one integrates this equation over the volume of the material or device being sampled, one gets:

$$\frac{\partial}{\partial t}\int_V \Delta n dV = \int_V \frac{\nabla \cdot \bar{J}_n}{q} dV + \int_V G dV - \frac{1}{\tau_n}\int_V \Delta n dV. \tag{4.19}$$

Or the total carrier density is:

$$\frac{\partial \Delta n_t(t)}{\partial t} = \int_V \frac{\nabla \cdot \bar{J}_n}{q} dV + G_n(t) - \frac{\Delta n_t(t)}{\tau_n}.$$

Here, n_t is the total number of excess electrons as a function of time.

For a confinement structure, the flow of electrons is blocked such that carrier flow out of the device is prohibited. Applying Gauss' law to the second term, one gets the divergence of J to vanish in confinement structures.

$$\int_V \nabla \cdot \bar{J}_n dV = \oint_S \bar{J} \cdot n dS = 0. \tag{4.20}$$

Therefore, the recombination dynamics in a confinement structure are:

$$\frac{\partial \Delta n_t}{\partial t} = G_n(t) - \frac{\Delta n_t(t)}{\tau_n}. \tag{4.21}$$

where G_n is the total number of electrons generated by an external source per $\mathrm{cm}^{-3}\,\mathrm{s}^{-1}$.

Thus, confinement structures simplify the direct measurement of carrier lifetime by either steady state and transient measurement methods by providing unambiguous data analysis. Dedicated confinement structures are made for the primary purpose of measuring

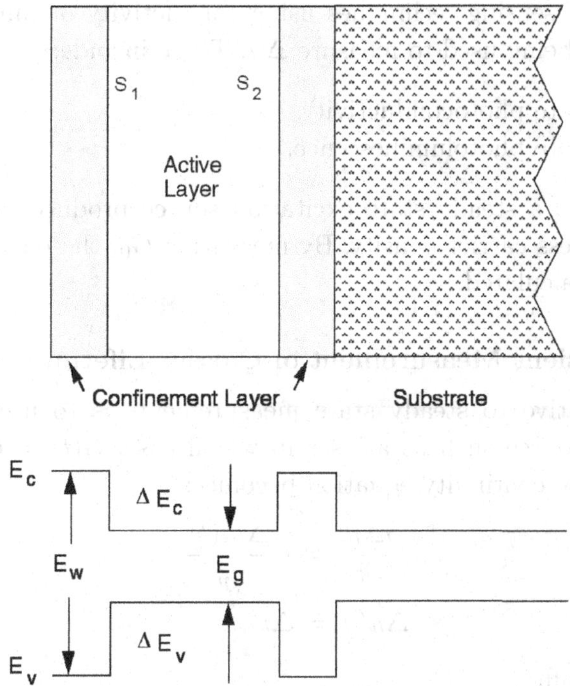

Fig. 4.2. Schematic of the band bending in a double heterostructure.

the carrier lifetime of a material that is being developed in the laboratory. The schematic in Fig. 4.2 is a typical example of a confinement structure made to measure carrier lifetime in thin film materials. A several hundred-micron thick wafer suffices as a confinement structure for materials such as silicon and germanium. These measurements on wafers greatly benefit from temporary or permanent surface passivation.

4.6. Steady State Measurement of Excess Carrier Density

For a steady-state measurements, Δn_t is constant under a constant source of carrier generation. This well-known result provides basis of steady-state measurements of excess electron (or carrier) density.

$$\Delta n_t = G_0 \tau_n. \tag{4.22}$$

A variety of sensing techniques using conductivity or luminescence sensors are being used to measure Δn. These include:

- Steady state photoconductivity.
- Steady state photoluminescence.

In each case, a steady state excitation source produces a constant value of excess carrier density. By measuring G_0, the recombination lifetime is calculated.

4.7. Transient Measurement of Carrier Lifetime

The alternative to steady state measurements is to use a pulsed excitation source such as a laser in which case, $G(t) = G_0\delta(t)$. In this case, the continuity equation becomes:

$$\frac{\partial \Delta n_t}{\partial t} = -\frac{\Delta n_t(t)}{\tau_n}$$
$$\Delta n(0) = \Delta n_0 \tag{4.23}$$

with the result:

$$\Delta n_t(t) = \Delta n_0 \exp(-t/\tau_n). \tag{4.24}$$

The excess carrier density versus time $\Delta n(t)$ is measured by means of transient photoconductivity, photoluminescence, or free-carrier absorption. The carrier lifetime is measured by fitting the transient decay data of the respective measurement. In the transient case, recombination occurs in the "dark"; i.e. $G_n = 0$ during the measurement of decay time. As stated earlier, Eq. (4.24) is valid for a confinement structure. When contacts or junctions are involved, current flows in and out of the active area. The current flow must be included in the total solution, and the solutions are generally more complicated.

4.7.1. *Photoconductivity*

External sources generated electron-hole pairs in a material or device, producing an additional component of conductivity. The most common laboratory light source is a monochromatic laser.

The photoconductivity is then written with the photo-generated carriers, Δn and Δp.

$$\begin{aligned}\sigma_n &= q\mu_n(n + \Delta n)\\ \sigma_p &= q\mu_p(p + \Delta p).\end{aligned} \tag{4.25}$$

The photoconductivity is obtained by subtracting the "dark" conductivity from the total conductivity. The measurement of $\Delta\sigma$ is convoluted with any mobility changes as μ is a function of Δn. Therefore, mobility must usually be accounted for in the data analysis.

$$\begin{aligned}\Delta\sigma(t) &= q\Delta n(t)\mu_n + q\Delta p(t)\mu_p\\ &= q\mu_n\Delta n(t)(1 + 1/b).\end{aligned} \tag{4.26}$$

Here, b is the mobility ratio μ_n/μ_p. The photoconductivity, $\Delta\sigma(t)$, is measured by either a steady-state or transient measurement technique. The dark or equilibrium conductivity is eliminated either by the measurement apparatus or by software. Then, the photoconductivity carriers produce a signal that is measured and contains lifetime information. We will discuss both steady-state and transient techniques in the following sections.

4.7.2. *Transient Photoconductive Decay*

A number of laboratory configurations provide a means to determine carrier lifetime by transient photoconductive decay. These basic techniques will be described in detail in a later section. The first data shown here will be obtained by a technique developed by the first author and coworkers at the National Renewable Energy Laboratory. This method is called resonant-coupled photoconductive decay (RCPCD), and will be described later in this work.[1,2] In these data, a pulsed laser is used to generate a population of excess carriers in the sample under test. A contactless sensor operating at 424 MHz is used to monitor the decay of the excess conductivity ($\Delta\sigma(t)$). The carrier lifetime is obtained by measuring the slope of

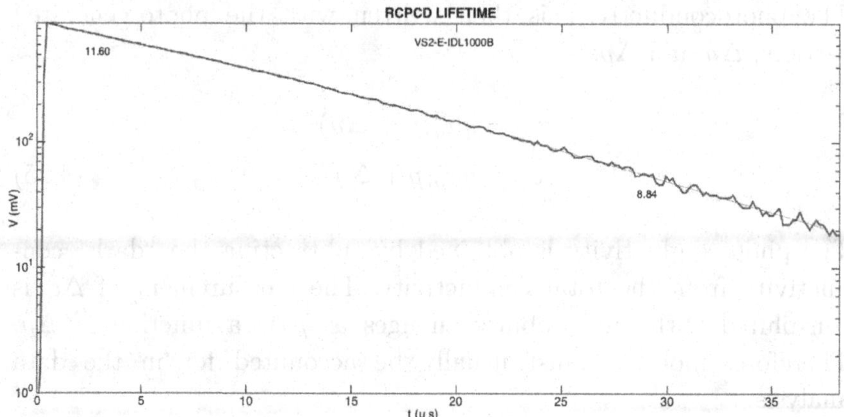

Fig. 4.3. Excess carrier decay in a silicon wafer that is about 300-μm-thick using the RCPCD technique measured at NREL.

$\Delta\sigma(t)$. Here the standard assumption is that the carrier mobility is constant over the measurement and the decay is purely controlled by $\Delta n(t)$. Models can be developed that include the mobility effects and that correction will be discussed in a later section.

Figure 4.3 above shows data from a commercial silicon wafer that is about 300 μm thick and doped *p*-type to a level of $2 \times 10^{14}\,\text{cm}^{-3}$. From the data, one sees that the initial, high-injection decay time is 22.8 μs. The excitation wavelength here is 1000 nm and is provided by an optical parametric oscillator (OPO) driven by a tripled YAG laser. The incident 1000 nm wavelength is weakly absorbed by the silicon wafer producing a nearly uniform excess carrier profile. The total photon flux is measured with a pulse energy meter and numerical estimates of injection level can be made.

4.8. Steady State Photoconductivity

When one multiplies the steady state equation by mobility, the result is:

$$\begin{aligned} q(\mu_n + \mu_p) * \Delta n &= G\tau q(\mu_n + \mu_p), \\ \Delta\sigma &= G\tau q(\mu_n + \mu_p). \end{aligned} \tag{4.27}$$

Therefore, the steady state lifetime is found by measuring $\Delta\sigma$ and G.

$$\tau = \frac{\Delta\sigma^* W}{qG(\mu_n + \mu_p)A}. \tag{4.28}$$

Experimentally, one usually measures the total photoconductivity, $\Delta\sigma_t$ of the illuminated portion of the wafer (area A). The wafer thickness, W, is measured and $\Delta\sigma$ calculated by inserting the sample dimensions: i.e.

$$\Delta\sigma_t = \frac{\Delta\sigma^* W}{A}. \tag{4.29}$$

Here, $\Delta\sigma_t$ is the total sample photoconductivity measured by the apparatus. This analysis will be described in a later section in connection with the quasi-static photoconductive decay technique (QSSPC).

4.8.1. *The Quasi-Steady State Photoconductivty (QSSPC) Technique*

A technique, based on this steady state photoconductivity, has become the standard for the silicon wafer industry. This method is called the quasi-steady state photoconductivity technique (QSSPC, Sinton Consulting), and uses a long duration (10 milliseconds) white flash lamp to produce a (nearly) steady state photoconductive signal in the sample.[3] The sample is placed in proximity to a small coil that is a component of a 10 MHz impedance bridge. The dark conductivity is cancelled by circuitry in the pickup coil circuit. The photoconductivity of the sample produces an induced current in the pickup coil that is processed and converted to a sample photoconductivity. A photodetector in the illuminated region measure the incident light flux as a function of time. The operator inputs to the processing computer include sample thickness, doping density, and surface reflectivity. Using the above equation and the samples parameters, the lifetime is calculated as a function of the incident light flux. The sample mobility is calculated using the well-known algorithms that use the doping density and injected carrier densities. The measuring system is calibrated by using a reference sample of known conductivity. The output of the computer display is the calculated QSSPC lifetime versus incident white light intensity.

The same silicon wafer, that was measured above by the photoconductive decay technique, was also measured by QSSPC. These data were generated by the QSSPC system at NREL. The surface is polished but not passivated. The intensity profile of the flashlamp excitation source is shown the dotted curve of Fig. 4.4(a). The QSSPC signal is shown by the solid curve of Fig. 4.4(a).

The data was measured by the author at NREL using a Sinton Instruments QSSPC apparatus. Figure 4.4(b) plots the lifetime versus calculated excess carrier density produced by the software analysis program of the instrument. The carrier lifetime varies from about 8 μs at 5×10^{14} to about 10 μs at 2.5×10^{15} cm^{-3}. The recombination lifetime over this range of injected carriers can be easily obtained from these data.

Both measurements (QSSPC and RCPCD) show the deep level saturation effects that were described in the previous section and modeled in a later section by the author at NREL. This saturation behavior is very common in most silicon wafer material, as the lifetime in the latter is primarily controlled by point or extended

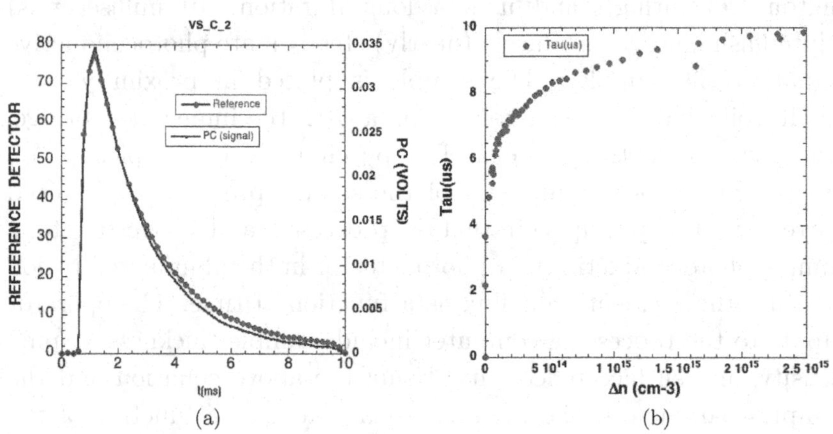

(a) (b)

Fig. 4.4. (a) The solid curve is the QSS photoconductive data with an overlay of the reference cell (dotted curve) data. (b) is the calculated lifetime using the reference cell data to calculate the instantaneous injection level.

defects. The quantative variation of lifetime with injection level is most clearly seen and recorded in the QSSPC measurement. The physics of deep level trapping and recombination will be expanded in Chapter 9.

When a volume is bounded by a high field region or junction, charge separation occurs and carriers diffuse out of the region. This process competes with recombination and the decay becomes more complex. Charge separation will be discussed in Chapter 12.

References

1. Ahrenkiel, R. K. (2013). *J. Vac. Sci. Technol.* **B31**, 04D113-1–04D113-8.
2. Ahrenkiel, R. K. and Johnston, S. W. (2003). *Materials Science and Engineering* **B102**, 161–172.
3. Sinton, R. A. and Cuevas, A. (1996). *Appl. Phys. Lett.* **69**(17), 2510–2512.

CHAPTER 5

Carrier Dynamics in Planar Device Structures

5.1. Fourier Mode Analysis

The majority of materials or devices to be tested are planar in structure. This designation characterizes everything from wafers to thin films. Therefore, a mathematical approach has been developed over the years to develop a model to be used for characterizing the excess carrier dynamics of these structures. The key input parameters in these structures are the material thickness and the respective SRVs at the front and back surfaces. Here, we will develop the Fourier modal analysis that has been widely adopted over many years to analyze heat flow in materials. The thickness of these planar structures typically vary from submicron (thin film) to several hundred microns (wafers).

One of the advantages of transient techniques is that unique decay curves are almost always found in the time evolution of excess-carrier decay. The functional characteristic of these decay curves contain information about the nature of the dominant recombination process. We will discuss the methods used to calculate the expected functional form of these curves. Also, we will apply theoretical procedures to analyze laboratory data.

5.2. Analysis of Transient Decay by the Fourier Method

5.2.1. *Devices of Finite Thickness*

The solution of the time-dependent diffusion equation usually starts with the following substitution:

$$\rho(x,t) \equiv U(x,t)\exp(-t/\tau) \tag{5.1}$$

This solution implies that there is a unique, bulk minority-carrier lifetime τ. On substitution into the diffusion equation, the exponential term $\exp(-t/\tau)$ cancels, and the differential equation involving $\rho(x,t)$ is transformed into the well-known equation developed for heat flow:

$$\frac{\partial U(x,t)}{\partial t} = D\frac{\partial^2 U(x,t)}{\partial x^2} \tag{5.2}$$

A number of researchers have used the time-dependent diffusion equation to analyze the transient PL decay in DH structures. The solutions to this boundary value equation subject for a variety of geometries are well documented in the heat flow literature.[1,2] This method of calculations have been carried over to the minority-carrier diffusion problem by Boulou and Bois[3] and others.[4,5] Boulou and Bois were among the first workers to calculate solutions to this problem and used the heat-flow mathematical techniques to analyze cathodoluminescence decay. Early calculations by van Opdorp and coworkers[6] used the time-dependent diffusion equation to calculate minority-carrier decay in structures with various geometries. Other formulations of the problem were given by 't Hooft and van Opdorp[7] and Ahrenkiel and Dunlavy.[5]

For symmetric structures, such as the double heterostructure, one may use the formalism developed by Carslaw and Jaeger.[2] The mathematics are developed for current flow in infinite planes of thickness d. An infinite Fourier series solution of Eq. (5.2) is.

$$U_n(x,t) = A(\cos\alpha_n x + b_n \sin\alpha_n x)e^{-\beta_n t} \tag{5.3}$$

where the total solution is:

$$U(x,t) = \sum_n U_n(x,t) \tag{5.4}$$

Here the parameters α_n are determined by the boundary conditions.

Substitution of $U(x,t)$ into Eq. (5.3) requires term by term equality producing a set of relationships between the α_n and β_n:

$$\beta_n = D\alpha_n^2 \tag{5.5}$$

The boundary conditions at $x = 0$ and $x = d$ equate the diffusion currents to the recombination currents at the two interfaces. Here S_1 and S_2 are the recombination velocities at the entrance and exit interface, respectively.

Therefore:

$$\begin{aligned} \left| qD\frac{\partial U_i}{\partial x} = qU_iS_1 \right|_{x-0}; \\ \left| qD\frac{\partial U_i}{\partial x} = -qU_iS_2 \right|_{x=d}. \end{aligned} \tag{5.6}$$

The boundary condition at the front surface above provides the relationship $b_n = S_1/(\alpha_n D)$ at $x = 0$. The boundary condition at $x = d$ produces an equation containing the recombination velocities S_1 and S_2 and contains the other unknown quantity α_n:

$$\tan \alpha_n d = \frac{\alpha_n d(S_1 + S_2)d/D}{(\alpha_n d)^2 - S_1S_2(d/D)^2} \tag{5.7}$$

A series of discrete values, α_n, are solutions to each term in the series; i.e. $n = 1, 2, 3$, etc. These values of α_n are then substituted back into Eqs. (5.3) and (5.4). It can be shown that the functions $U_n(x,t)$ form an orthogonal basis set for describing $\rho(x,t)$. These functions $U_n(x,t)$ are mathematically similar to the wave functions used in quantum mechanical problems.

We define the set of values α_n as then eigenvalues of Eq. (5.7) and the latter is called an eigenvalue equation. We will also use the term double heterostructure (DH) to define the planar structure as has become the standard designation for thin film epitaxial structures.

One can apply the asymmetric case, $S_1 \neq S_2$ to describe junction devices, such as a p-n device. For example, one side could be a depletion region and the second a contact or carrier blocking layer, commonly called a back surface field in photovoltaic devices. We will refer to this version of the eigenvalue equation as asymmetric equation and the device as an ADH. One can define dimensionless recombination velocities:

$$Z_1 = \frac{S_1 d}{D};$$
$$Z_2 = \frac{S_2 d}{D}. \tag{5.8}$$

We will also define the quantity $\theta_n \equiv \alpha_n d$. The eigenvalue equation is then written as:

$$\tan(\theta_n) = \frac{\theta_n(Z_1 + Z_2)}{\theta_n^2 - Z_1 Z_2}. \tag{5.9}$$

For the case of equal interface recombination velocities, (or $S_1 = S_2 = S$), the equation simplifies to:

$$\tan(\alpha_n d) = \frac{2(\alpha_n d)(Sd/D)}{(\alpha_n d)^2 - (Sd/D)^2} \tag{5.10}$$

One can write this equation in more compact form by defining the dimensionless quantities $\theta_n = \alpha_n d$ and $z = Sd/D$. In terms of these new parameters, the nth eigenvalue equation is:

$$\tan(\theta_n) = \frac{2z\theta_n}{\theta_n^2 - z^2};$$
$$\theta_n = \alpha_n d; \tag{5.11}$$
$$z = \frac{Sd}{D}.$$

The solutions, $U_n(x,t)$, of this transcendental equation are found by commercial nonlinear equation solving routines. The eigenvalues or roots, $\theta_{n(5.11)}$, of the equation are seen by plotting the two sides of Eq. (5.10) versus θ/π and finding the intersection points. Figure 5.1 shows a plot of tan θ (Curve B) and the function $2z\theta/(\theta^2 - z^2)$, (Curve A) with the points of intersection indicated by a solid circle.

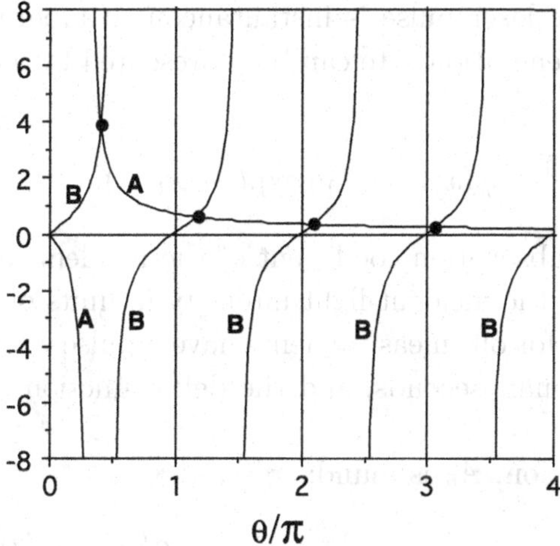

Fig. 5.1. Graphical representation of the solutions to the transcendental Eq. (5.11).

The equation is plotted above with $z = 1$ and the intersection points are indicated. The function tan (θ) intersects with the hyperbolic function once in each quadrant, i.e. between 0 and π, π and 2π, and $n\pi$ and $(n+1)\pi$, etc. These intersections provide the unique solutions (θ_n) that are eigenvalues of Eq. (5.11). For this example of $z = 1$, the value of θ_1 is 0.30 π. As can be seen from Fig. 5.1, the higher order solutions $(n > 1)$ are $\theta_n \sim (n-1)\pi$ where n is the index number. In the limit of $S = 0$, the eigenvalues for $n = 1, 2, 3$, etc.:

$$\begin{aligned} \theta_n &= \alpha_n d = 0, \pi, 2\pi, \ldots \\ \beta_n &= 0, \frac{D\pi^2}{d^2}, \frac{4D\pi^2}{d^2}, \ldots \end{aligned} \tag{5.12}$$

For the general case, the solutions can be solved by any of the commercial nonlinear equation solver packages. We used the Matlab routine called fsolve to the nonlinear equation. Here, θ_0 (theta0) is cycled through each quadrant (n-value) as a starting value and the solution is accurately and quickly calculated by iteration to the desired accuracy for that particular quadrant.

The optical excitation produces an initial excess minority-carrier density given by Beers' law. If one assumes that the duration

(FWHM) of the laser pulse is instantaneous compared to diffusion processes, the generation rate can be represented by a delta function $\delta(t)$. Therefore:

$$\rho(x, 0) = I_0 \alpha \exp(-\alpha x)\delta(t). \tag{5.13}$$

where α is the absorption coefficient of the incident monochromatic light. Here I_0 is the incident light intensity in units of photons/cm^2. The lasers used for our measurements have a pulse widths of picoseconds to several nanoseconds, and the delta function approximation is adequate.

By substitution, A_n is found:

$$A_n = I_0 \frac{2\alpha_n^2}{\left(\alpha_n^2 + \left(\frac{S}{D}\right)^2 d + 2\frac{S}{D}\right)} \frac{\alpha(\alpha + S/D)}{(\alpha_n^2 + \alpha^2)}. \tag{5.14}$$

Finally, one finds $\rho(x, t)$ equal to:

$$\rho(x, t) = \exp(-t/\tau) \sum_{n=1}^{\infty} \exp(-\beta_n t) A_n \left[\cos(\alpha_n x) + \frac{S}{\alpha_n D} \sin(\alpha_n x)\right]. \tag{5.15}$$

In low injection, $\rho(t)$ is the integral of $\rho(x, t)$ over the active volume, or from $x = 0$ to $x = d$.

$$\rho(t) = \sum_{n=1}^{\infty} A_n C_n \exp(-t/\tau_n). \tag{5.16}$$

Here τ_n is the decay time of the nth mode, which is given by:

$$\frac{1}{\tau_n} = \frac{1}{\tau} + \beta_n = \frac{1}{\tau} + D\alpha_n^2. \tag{5.17}$$

The asymptotic or dominant term after diffusion has flattened the profile is:

$$\frac{1}{\tau} \approx \frac{1}{\tau_B} + \beta_1 = \frac{1}{\tau_B} + D\alpha_1^2 \cdot \frac{1}{\tau} \approx \frac{1}{\tau_B} + \beta_1 = \frac{1}{\tau_B} + D\alpha_1^2. \tag{5.18}$$

Here, α_1 is found from the solution to the eigenvalue equation.

Also, C_n comes from the integration of $\rho(x)$ over the active region that has a thickness d:

$$C_n = 1/\alpha_n \left[\sin \alpha_n d + \frac{S}{\alpha_n D}(1 - \cos \alpha_n d)\right]. \tag{5.19}$$

The function $\rho(t)$ is the sum of a series of exponential decays of the individual modes. Each mode has a specific time constant $\beta_n + 1/\tau$ and is weighed by a coefficient $A_n \cdot C_n$. Thus, the general solution for the total decay of $\rho(t)$ is nonexponential and only an instantaneous lifetime can be defined. The coefficients β_n increase with n and therefore $\beta_1 < \beta_2 < \beta_3$, etc.

A. Low recombination velocity: $z < 1$

For $S = 0$, Eq. (5.19) shows the following difference between successive values of β_n:

$$\beta_{n+1} - \beta_n = (2n - 1)\frac{D\pi^2}{d^2}. \tag{5.20}$$

The diffusion transit time τ_D is defined as:

$$\tau_\mathrm{D} = \frac{d^2}{D\pi^2}. \tag{5.21}$$

or,

$$\beta_{n+1} - \beta_n = \frac{(2n - 1)}{\tau_\mathrm{D}}. \tag{5.22}$$

At $z = 0$, the values of β are $\beta_1 = 0$, $\beta_2 = 1/\tau_\mathrm{D}$, $\beta_3 = 4/\tau_\mathrm{D}$, $\beta_4 = 9/\tau_\mathrm{D}$, etc. The first mode dominates the decay rate at long times. The higher modes decay very rapidly with time as β_n increases rapidly with the index n. For nonzero, but small values of z, the first ($n = 1$) mode dominates at times longer than τ_D. At longer times (i.e., $t > \tau_\mathrm{D}$), $\rho(t)$ is approximated by a single exponential decay process, when the first decay mode dominates.

$$I_\mathrm{PL}(t) \cong \frac{1}{\tau_\mathrm{R}} A_1 C_1 e^{-\beta_1 t} e^{-t/\tau} \tag{5.23}$$

The long-term lifetime is clearly $1/(\beta_1 + 1/\tau)$. If $z = 0$, then $\beta_1 = 0$ and the excess density decays with the bulk lifetime τ. The term β_1 is related to the surface/interface recombination rate and the SRV is contained in β_1.

$$\frac{1}{\tau_S} = D\alpha_1^2. \tag{5.24}$$

The range of values of z for heterointerfaces in DHs is less than unity using current growth techniques. Equation (5.7) may be solved for θ as a function of z with a nonlinear equation solver.

Figure 5.2 is a calculation of the excess carrier density in a model system that represents a high quality, float zone grown silicon wafer with a bulk lifetime of 1 ms. The surfaces are assumed to be passivated with one of the standard techniques such as a thermal SiO_2 oxide or iodine/methanol solution. The SRV is taken to be 100 cm/s which is near the upper limit for such passivation

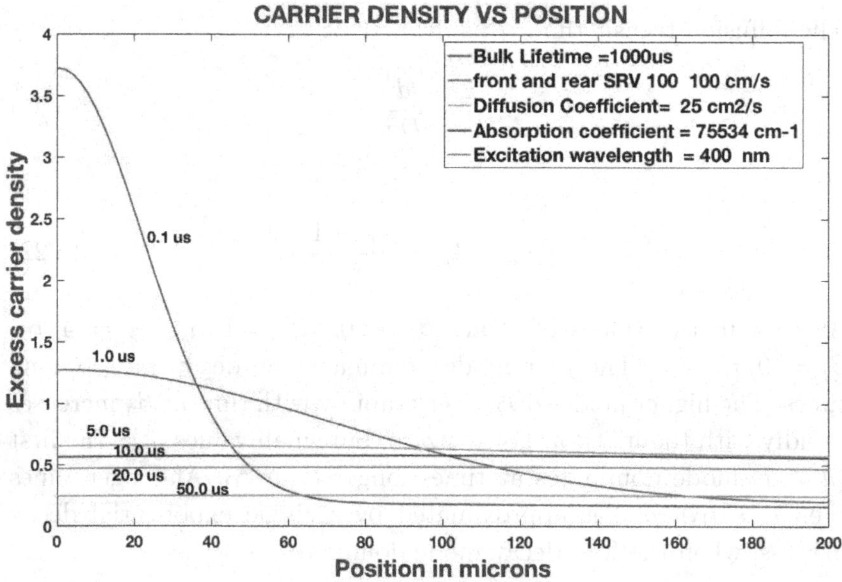

Fig. 5.2. Numerical calculation of the excess carrier concentration in a 200-mm thick silicon wafer after a pulse-laser excitation of 400 nm light.

techniques. The calculation assumes an incident impulse excitation at a wavelength of 400 nm. At $t = 0.1\,\mu$s, the excess carriers have diffused out of the initial Beers law profile ($1/\alpha \sim 0.1\,\mu$m) and at $t = 1\,\mu$s, carriers have started to accumulate at the rear surface. At $t = 10\,\mu$s, the excess carrier density has become "flat" owing to diffusion. The flat profile then decays to zero dominated by bulk recombination and diffusion to the surface from the flat profile.

B. Approximate solution to Lifetime for Thick Films

Figure 5.3, Curve A, comes from the exact, numerical solution to Eq. (5.10). Curve A plots the exact numerical solution for τ_S versus S over the range $S = 1$ to 5×10^3 cm/s. The parameters here are typical for a commercial silicon wafer with a thickness of 200 μm and a carrier mobility of 1000 cm^2/Vs.

Boulou and Bois[8] provided an approximate solution to Eq. (5.25) that is accurate over a wide range of S. This formula for the surface lifetime is:

$$\tau_S = \frac{d^2}{\pi^2 D} + \frac{d}{2S} \tag{5.25}$$

By rearrangement, one may define the dimensionless surface recombination lifetime, τ_S, as a function of z:

$$\frac{d^2}{D\tau_S} = \frac{2\pi^2 z}{\pi^2 + 2z} \tag{5.26}$$

The Boulou-Bois approximation, is plotted in as dashed line and is identical in this case with the exact solution (solid line), except for $S < 5$ cm/s.

For thick wafers, the diffusion of excess carriers to the surface becomes a significant factor in the recombination velocity. As the materials becomes thicker (i.e. d), the effect of the surface recombination becomes smaller. For ingots or very thick wafers, the surface becomes insignificant if the excitation source is able to generate the majority of carriers deeply in the material.

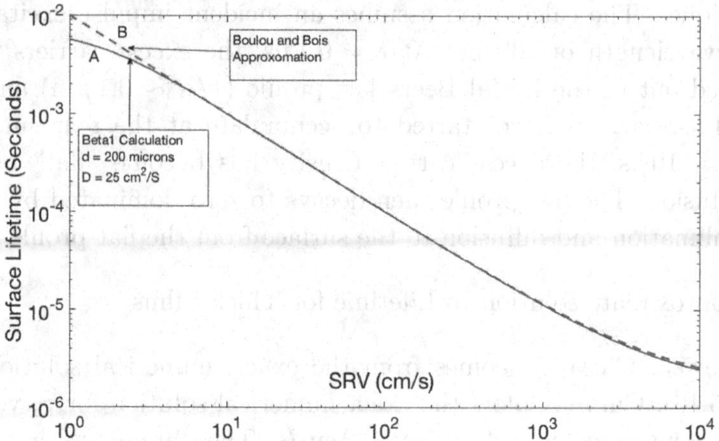

Fig. 5.3. Curve A: solid curve; the exact solution to Eq. (5.11). Curve B: dashed curve: the Boulou-Bois approximation.

The carrier lifetime for thins DH structures ($z < 1$) limit is given by:

$$\frac{1}{\tau} = \frac{1}{\tau_B} + \frac{2S}{d} \tag{5.27}$$

where S in the recombination velocity of the two interfaces and τ_B is the bulk lifetime. If the values of S are different for each surface, the total lifetime for thin films is:

$$\frac{1}{\tau} = \frac{1}{\tau_B} + \frac{S_1}{d} + \frac{S_2}{d}. \tag{5.28}$$

C. High Recombination Velocity: $z > 1$

At large values of z, θ approaches the value π as will be shown here. For very large values of $z (z \gg 1)$, the hyperbolic function is a small, negative number for $\theta < z$, and therefore the intersection points are near the zeros of $\tan\theta$. The first solution θ_1 is slightly less than π, θ_2 is slightly less than 2π, etc. For z approaching infinity, the solutions of θ_n are the zeros of $\tan\theta$. These solutions are for $n = 1, 2, 3$, etc.:

$$\begin{aligned} \theta_n &= \alpha_n d = \pi, 2\pi, 3\pi \ldots \\ \beta_n &= D\pi^2/d^2, 4D\pi^2/d^2, \ldots \end{aligned} \tag{5.29}$$

Thus β_n may be written as $(\frac{n^2}{\tau_D})$ and the difference of successive terms goes as:

$$\beta_{n+1} - \beta_n = \frac{(2n+1)}{\tau_D} \tag{5.30}$$

The successive values of β increase as $\beta_1 = 3/\tau_D$, $\beta_2 = 5/\tau_D$, $\beta_3 = 7/\tau_D$, etc., and the higher modes do not decay as quickly as for the low z case. Longer times are required for the first mode (i.e., $n = 1$ term) to become dominant.

For intermediate values of z, Eq. (5.10) may be solved numerically to obtain the eigenvalue θ_n. The exact solutions for the θ_n are obtained using standard numerical procedures for the solution of nonlinear or transcendental equations. Alternatively, the Boulou and Bois approximation Eq. (5.25) may be used.

The expression for $\rho(t)$ at long times ($t > \tau_D$) is now given by:

$$\rho(t) \approx A_1 C_1 \exp(-t/\tau) \exp\left(\frac{-t}{\frac{d^2}{\pi^2 D} + \frac{d}{2S}}\right) \tag{5.31}$$

For the case of $z \gg 1$, the minority-carrier lifetime is an exponential function of the minority-carrier diffusivity. The PL lifetime is limited by the diffusion rate to the interfaces. From Eq. (5.13), one can write the total carrier lifetime as:

$$\frac{1}{\tau} = \frac{1}{\tau_B} + \frac{D\pi^2}{d^2} = \frac{1}{\tau_B} + \frac{1}{\tau_D}; \quad (z >> 1). \tag{5.32}$$

In this limit, the lifetime is controlled by the diffusion transit time to the interfaces and independent of the SRV. A model calculation of the excess carrier decay kinetics for a silicon wafer is shown in Fig. 5.4. This model calculation uses the Fourier mode method for a silicon wafer of thickness 250 microns. The excitation wavelength is fixed at 532 nm and the variable here is the surface recombination velocity (S). The calculation shows that there are dramatic changes in the decay curve for $S < 10^3$ cm/s and then the changes are fairly small for larger values of S.

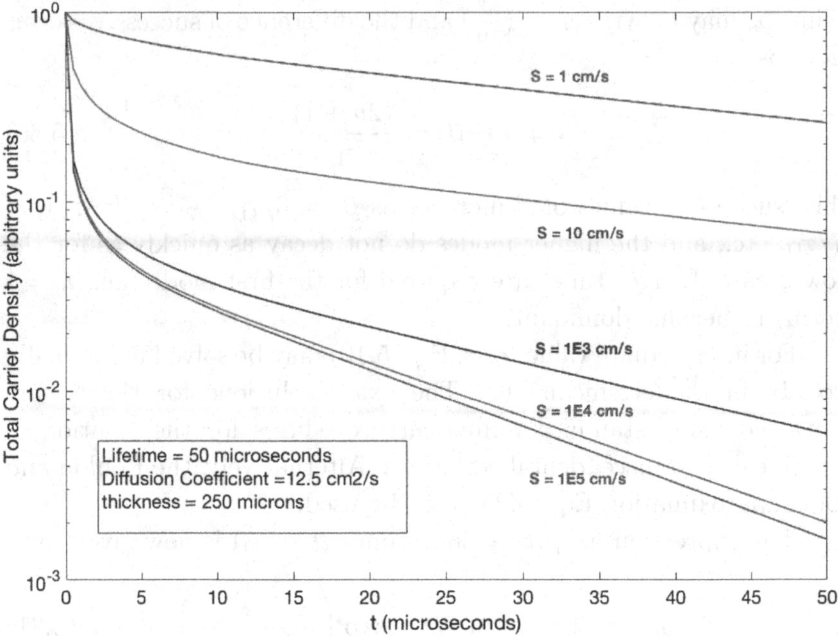

Fig. 5.4. Model calculation for the total excess carrier density versus time for a 250-mm thick silicon wafer with the parameters shown. The excitation wavelength is 532 nm.

For smaller values of S (or z), the long term decay approaches an asymptotic value that depends on the bulk lifetime, sample thickness, diffusivity and S. As the wafer thickness becomes very large, the asymptotic lifetime is simply the bulk lifetime. In the very large SRV range, the asymptotic lifetime is:

$$\begin{aligned}\rho(t) &\approx A_1 C_1 \exp(-t/\tau) \exp\left(-t\frac{\pi^2 D}{d^2}\right) \rho(t) \\ &\approx A_1 C_1 \exp(-t/\tau) \exp\left(-t\frac{\pi^2 D}{d^2}\right). \end{aligned} \tag{5.33}$$

This is the case for the model calculation for $S > 1$ E4 cm/s.

In measuring the SRV for thin films, a standard procedure is to make two structures. These are identically processed except for the active-layer thickness d. From these two data sets, one can obtain unique values of bulk lifetime and S. This information is, of course, very useful for process monitoring and development.

For wafers and thicker structures, the surface recombination values can be determined by used pulsed excitation and varying the excitation wavelength. As the wavelength becomes shorter (and the absorption coefficient larger), the initial decay becomes steeper as shown in the model calculation below. In this model, the sample is silicon and some typical parameters of unpassivated wafers are used. The absorption coefficient of silicon varies by orders of magnitude as the excitation energy varies from the direct bandgap to the indirect bandgap wavelength range. A model calculation is shown in shown in Fig. 5.5.

Here the ordinate represents the excess carrier density, $\rho(x,t)$, as measured by any transient technique. When the surface recombination parameter, S, is less than 10 cm/s, there is no detectable difference in the shape of the decay curves.

The data from photoconductive decay measurements on an unpassivated silicon wafer are shown in Fig. 5.6. The wafer is taken

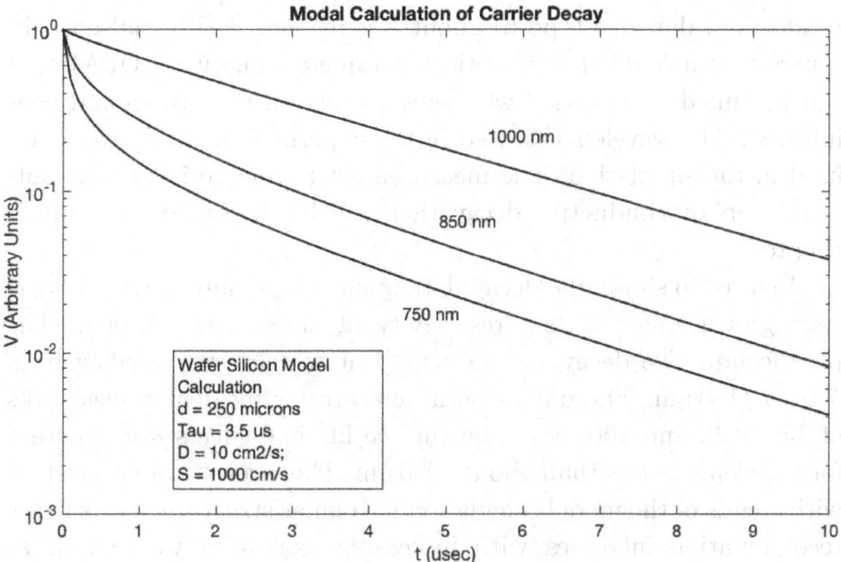

Fig. 5.5. The parameters used in this calculation are shown in the figure. The SRV is fixed at 1000 cm/s and excitation wavelength is varied.

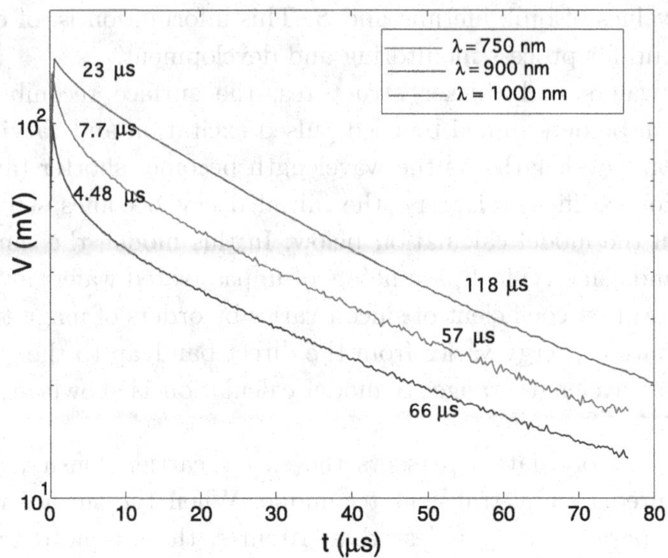

Fig. 5.6. The photoconductive decay data of an unpassivated silicon wafer showing the three wavelengths used for pulsed excitation.

from a batch of commercially available material purchased from a vendor and doped p-type to about $5 \times 10^{15}\,\text{cm}^{-3}$. The pulsed light source was a YAG driven optical parameter amplifier (OPA) that can be tuned over a wide wavelength range in the visible and near infrared. The wavelengths used in the experimental study were also used in the simulation. The measurement technique is by resonant-coupled photoconductive decay that will be described in the next chapter.

Figure 5.6 shows the decay data of an n-type, unpassivated float-zone grown wafer with a resisitivity of about 1000 ohm-cm. The photoconductive decay was measured at wavelengths varying from 500 to 1100 nm. The data shown here are for incident wavelengths of 750, 900, and 1000 nm. The surface lifetime effects are apparent for wavelengths less than about 1000 nm. These results are consistent with trends of the model calculation for planar structure. The surface recombination increases with increasing excitation wavelength as predicted. These data of will be fit with the modal function of Eq. (5.4) using a nonlinear least squares technique. This curve fitting procedure provides values of S, τ, and D.

To verify that surface recombination effects are producing these decay curve variation, the samples were rerun using a iodine/methanol passivation technique. Figure 5.6 shows the same sample after an HF etch while being immersed in iodine-methanol solution. The same excitation wavelengths of 800, 900, and 1000 nm were used for the measurement. The decay functions are nearly identical and are independent of the excitation wavelength. The data were not normalized for the wavelength dependence of the excitation pulse energy.

The slope, and hence the lifetime, is not constant over the entire range of injection level. The behavior is indicative of deep level filling and the subsequent nonlinear Shockley-Read-Hall recombination rates were described earlier in this work. In summary, the recombination in the passivating solutions can be described in terms of a bulk SRH mechanism with the surface effects being insignificant. The data is measured in a passivating iodine/methanol solution and shown in Fig. 5.7.

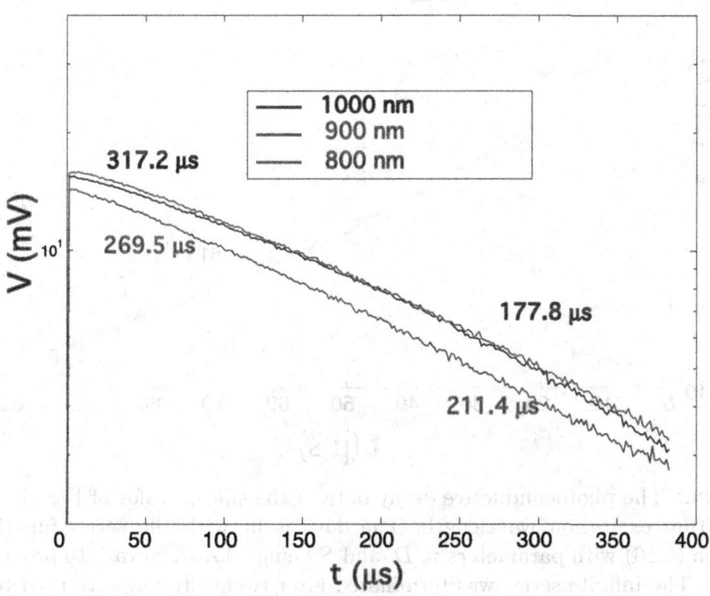

Fig. 5.7. The photoconductive decay of the wafer of Fig. 5.6 but with the wafer immersed in passivating iodine/methanol solution.

We see an unchanging decay function as seen in the data as the source is tuned from 800 nm to 1000 nm. The nearly constant decay function is indicative of insignificant surface recombination.

Figure 5.8 shows a nonlinear least squares fit of the 900 nm data for the unpassivated case. A similar S-value was found for the 800 nm data. The decay curve was analyzed using 1000 points of digitized data. The data fit produced the S, τ, and D values that are shown in the figure. The transient data are fit with a bulk lifetime value of 61.1 μs as shown in Fig. 5.8. We notice that the asymptotic decay time is much less than the similar time in the passivated case. This lifetime drops from the bulk, low-injection value of about 200 μs to about 60 μs. The effective lifetime is reduced by the addition of the surface component to the bulk component as predicted by Eq. (5.24). Using the expression $1/(\beta_1 + 1/\tau)$ for the asymptotic decay rate and

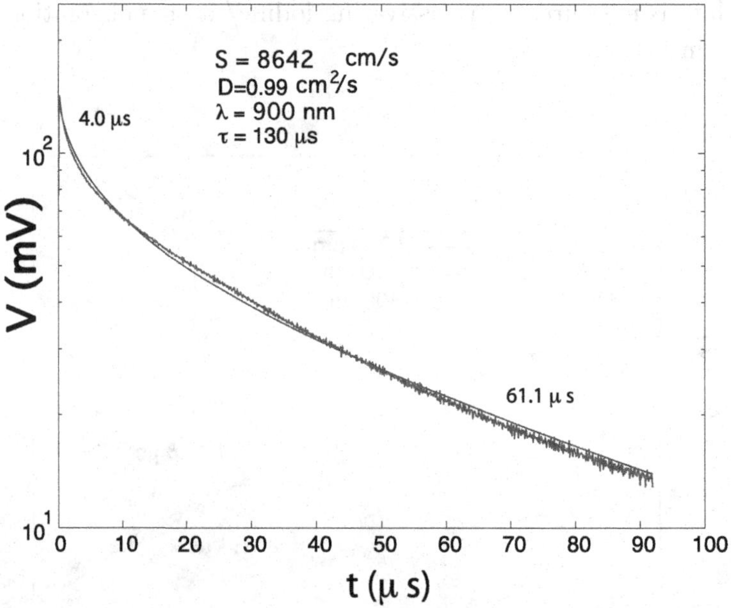

Fig. 5.8. The photoconductive decay data of the silicon wafer of Fig. 5.6 using the 800 nm excitation wavelength. The data is fit with the series function of equation (5.16) with parameters τ, D, and S being allowed to vary to provide the best fit. The infinite series was terminated after twenty terms were used to give the fit shown. The fit provides an SRV value of 8642 cm/s.

assuming that S the fitted value, one can calculate the diffusivity D;

$$\frac{1}{\tau} \approx + \frac{1}{\tau_{\mathrm{B}}} + \left(\frac{1}{\frac{d^2}{\pi^2 D} + \frac{d}{2S}} \right) \tag{5.34}$$

The excitation at 1000 nm shows a very weak surface recombination effect. Here the effective lifetime is approximately given by Eq. (5.33).

5.3. Recombination Kinetics in Semi-Infinite Materials

Here we will analyze the excess carrier decay in a sample with infinite thickness, so that the recombination at a back surface does not alter the measured lifetime.

As before, the initial distribution depends strongly on the wavelength of the monochromatic light. After the nearly instantaneous generation of electron-hole pairs at $t = 0$, the one-dimensional continuity equation can be written,

$$D_n \frac{d^2 \rho(x,t)}{dx^2} = \frac{d\rho(x,t)}{dt} + \frac{\rho(x,t)}{\tau} - \alpha I_0 * \exp(-\alpha x)\delta(t) \tag{5.35}$$

Defining $U(x,t)$ as $\rho(x,t) * \exp(t/\tau)$, the third term is eliminated.

$$D_n \frac{d^2 U(x,t)}{dx^2} = \frac{dU(x,t)}{dt} - \alpha I_0 \exp(t/\tau) * \exp(-\alpha x)\delta(t) \tag{5.36}$$

We will now perform a Laplace transform of this equation, from the time domain to the Laplace frequency space. The Laplace transform is defined as:

$$U(x,s) \equiv \int_0^\infty \exp(-st) * U(x,t) dt. \tag{5.37}$$

Here, s is the Laplace transform frequency variable. The transform of Eq. (5.36) is:

$$D_n \frac{d^2 U(x,s)}{dx^2} = sU(x,s) - \alpha I_0 \exp(-\alpha x). \tag{5.38}$$

This equation is solved in the x-domain with the solution:

$$U(x,s) = A\exp\left(-\sqrt{\frac{s}{D_n}}x\right) + B\exp\left(+\sqrt{\frac{s}{D_n}}x\right) + \frac{\alpha I_0}{s-\alpha^2 D_n}\exp(-\alpha x). \quad (5.39)$$

As the materials is assumed to be infinitely thick, B must be zero. The coefficient A is calculated from the surface boundary condition:

$$D_n\left[\frac{dU}{dx}\right]_{x=0} = S_n U(0,s). \quad (5.40)$$

Here S_n is the surface recombination velocity at the surface where the light is incident. The final solution in Laplace transform space is:

$$U(x,s) = \frac{\alpha I_0}{s-\alpha^2 D_n}\left[\exp(-\alpha x) - \frac{S_n+\alpha D_n}{S_n+\sqrt{sD_n}}\exp\left(-\sqrt{\frac{s}{D_n}}x\right)\right]. \quad (5.41)$$

The total excess charge density is obtained by integrating $U(x,s)$ over all space.

$$U_t(x,s) = \int_0^\infty U(x,s)dx = \frac{I_0}{s-\alpha^2 D_n}\left(1 - \frac{\alpha\left(S_n+\alpha D_n\right)}{S_n+\sqrt{sD_n}}\right). \quad (5.42)$$

The total charge density as a function of time is given by the inverse transform of the above.

$$\rho_t(t) = \exp(-t/\tau) * L^{-1}\left[U_t(x,s)\right]. \quad (5.43)$$

$$\rho_t(t) = I_0\exp(-t/\tau)\left[\frac{S_n}{S_n-\alpha D_n}\exp(\alpha^2 D_n t)erfc(\alpha\sqrt{D_n t}) - \frac{\alpha D_n}{S_n-\alpha D_n}\exp\left(\frac{S_n^2 t}{D_n}\right)erfc\left(S_n\sqrt{\frac{t}{D_n}}\right)\right]. \quad (5.44)$$

In the limit of $S = 0$, the decay is a pure exponential with the bulk lifetime, as expected.

$$\rho_t(t) = I_0\exp(-t/\tau). \quad (5.45)$$

A sample calculation is shown here with the assumption of a direct bandgap material such as GaAs, with S_n as a parameter. The transient decay time shows a dramatic decrease at $t = 0$ with increasing S. The origin of this increased decay rate is that the carriers are initially in a Beer's law distribution near the entrance surface and interact strongly with the surface states. As the surviving carriers diffuse into the bulk and away from the surface, the recombination rate decreases dramatically. This process is also sensitive to the diffusion coefficient, D_n, and becomes more rapid as the latter increases.

The arguments in the calculation of exponential and error functions may become quite large with the large values of α and S_n that may commonly occur. For very large arguments, one may use a series expansion with high accuracy.

$$\exp(x^2) * erfc(x) \simeq \frac{1}{x\sqrt{\pi}}\left[1 - \sum_{1}^{\infty} \frac{(2j-1)!!}{(-2x^2)^j}\right]; \quad x \gg 10 \tag{5.46}$$

This expansion is quite accurate for value of the argument, x, at values large than about 10. In the above equation, the argument x is that in the $erfc(x)$ term.

When the absorption coefficient, α, is quite low, the reduction of the initial lifetime is quite small and becomes unobservable. In the limit of αD_n being much smaller the S_n, the effective ($t = 0$) decay time becomes:

$$\tau_{eff} = \tau\left(1 - \frac{\alpha D_n}{S_n}\right). \tag{5.47}$$

For moderate values of α and S_n, the initial decay time at $t = 0$ is:

$$\tau_{eff} = \frac{\tau}{1 + \alpha S_n \tau}. \tag{5.48}$$

By combining the bulk lifetime, τ, from the asymptotic decay time, with the initial slope, one calculate an S-value from the transient measurement.

In the limit of very high surface recombination ($S_n \gg \alpha D_n$), the decay times varies as:

$$\rho_t(t) = I_0 \exp(-t/\tau) \exp(\alpha^2 D_n t) erfc(\alpha\sqrt{D_n t}). \quad (5.49)$$

In this case, the dominant parameter is the diffusion coefficient.

A transient method of carrier decay is quite easily applied to crystalline silicon using a pulsed YAG laser as the light source. Here one uses the fundamental lasing wavelength ($\lambda = 1064\,\text{nm}$, $\alpha = 5.49$) and the first harmonic or doubled wavelength ($\lambda = 532\,\text{nm}$, $\alpha = 1.13 \times 10^4$) to make a transient decay measurement.

In silicon, absorption at 532 nm is a direct transition, whereas 1064 nm is in the indirect absorption region. This transition produces the orders of magnitude change in absorption coefficient. In silicon, the preferred measurement technique is photoconductive decay, as the indirect bandgap character of silicon provides relatively weak photoluminescence (for TRPL to be applicable). In Fig. 5.9, a calculation is presented of the simulated PCD decay measurement.

Here, the calculated decay time at $t = 0$ is 9 ns. One may use this combination of two wavelengths to provide a combination of

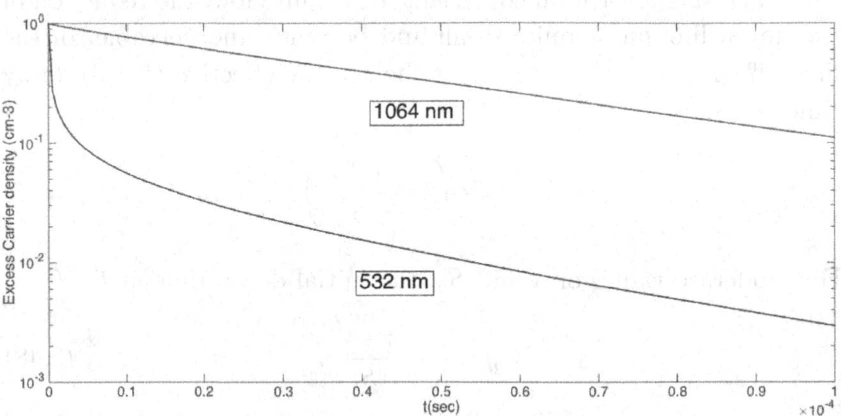

Fig. 5.9. Calculated photoconductive decay of an infinitely thick wafer of silicon with parameters: $\tau = 50\,\mu s$; $D = 12.5\,\text{cm}^2/\text{s}$, and the front surface $S_n = 1 \times 10^4\,\text{cm/s}$.

weak and strong absorption, to measure the surface recombination velocity and bulk lifetime in thick material.

5.4. Direct Bandgap Materials

For silicon, the standard method to measure transient carrier decay uses photoconductivity as the probe as the photoluminescence signal is very weak because of the indirect bandgap. For silicon and germanium, there are a large ranges of wavelengths available as the excitation sources can be tuned to excite the direct and indirect bandgap absorption regions. For direct bandgap materials, the range of wavelengths is limited as the absorption coefficient increases quickly from the absorption edge to over $1 \times 10^4\,\text{cm}^4$ for higher photon energies.

Kuciauskas[9] and coworkers developed a novel method to address this problem by using two photon absorption at sub-bandgap excitation wavelengths. In this case, the two-photon absorption coefficient

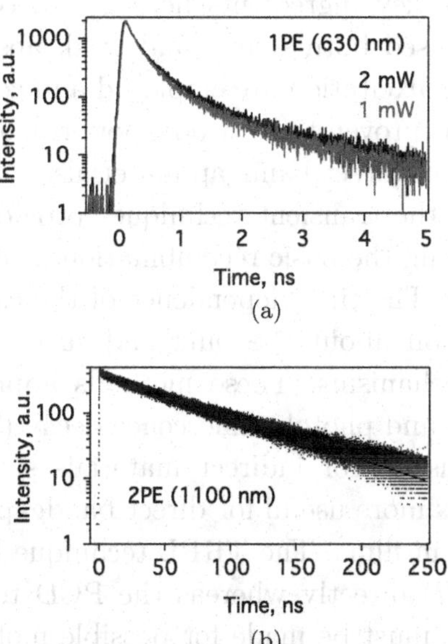

Fig. 5.10. (a) One photon absorption of crystalline CdTE. (b) Fig. 5.10 B: Two-photon absorption of the same crystal of CdTe.

is quite small and it is proportional to the square of the incident optical pulse. The surface recombination velocity is then obtained by comparing the PL decay with a higher energy, single photon PL decay with a two-photon absorption process. This technique was successfully applied to both crystalline CdTe and GaAs.

The data in Figs. 5.10(a) and (b) were obtained by a sequence of one-photon and two-photon TRPL measurements on a CdTe wafer by Kuciauskas and coworkers. The sample here was about 1-mm-thick and was nominally undoped. The background doping is about $1 \times 10^{14}\,\mathrm{cm}^{-3}$ and is *p*-type. The one-photon TRPL (curve a) shows the steep, initial decay characteristic of strongly absorbed light, followed by surface recombination. The two-photon data (curve b) used and excitation wavelength of 1100 nm (1.12 eV) and shows a bulk-dominated lifetime. From these data, the authors calculated a SRV value of $1.4 \times 10^4\,\mathrm{cm/s}$ and a bulk lifetime of 66 ns.

In summary, the transient techniques are a quick and convenient method to determine surface recombination velocity. Either photoconductive decay or photoluminescence decay are applicable to this measurement. The key ingredients here are to compare excitation wavelengths that used both strong and weak absorption such that surface and volume excitation are employed and compared. The two-photon method has proven to be a very powerful addition of tools to extricate the SRV of direct bandgap materials.

In conclusion, the transient techniques provide a very powerful method for analyzing the basic recombination mechanisms in a wide range of materials. The time dependence of the excess carrier decay contains information about the bulk and surface contributions to recombination mechanisms. These methods apply to both photoconductive (PCD) and photoluminescence decay (TRPL). The PCD method is most useful for indirect materials such as silicon. The TRPL technique is more useful for direct bandgap materials such as III-V and II-IV thin films. The TRPL technique has the advantage of measuring $\Delta n(t)$ directly whereas the PCD techniques measure $\Delta\sigma(t)$. Allowances must be made for possible mobility variations in the PCD measurement.

References

1. Crank, J. (1975). In *The Mathematics of Diffusion*, 2nd Ed., Clarendon, Oxford.
2. Carslaw, H. S. and Jaeger, J. C. (1978). In *Conduction of Heat in Solids*, 2nd Ed. Oxford University Press, Oxford.
3. Boulou, M. and Bois, D. (1977). *J. Appl. Phys.* **48**, 4713.
4. 't Hooft, G. W. and van Opdorp, C. (1983). *J. Appl. Phys. Lett.* **42**, 813.
5. Ahrenkiel, R. K. and Dunlavy, D. J. (1989). *J. Vac. Sci. Technol.* **A7**, 822.
6. van Opdorp, C., Vink, A. T., and Werkhoven, C. (1977). *Instr. Phys. Conf. Ser.*, No. 33b, p. 317.
7. t Hooft, G. W. and van Opdorp, C. (1986). *J. Appl. Phys.* **60**, 1065.
8. Boulou, M. and Bois, D. (1977). *J. Appl. Phys.* **48**, 4713.
9. Kuciauskas, D., Kanevce, A., Burst, J. M., Duenow, J. N., Dhere, R., Albin, D. S., Levi, D. H., and Ahrenkiel, R. K. (2013). *IEEE J. Photovoltaics* **3**, 1319.

CHAPTER 6

Transient Photoconductivity

6.1. Introduction

We have analyzed many of the common steady state techniques. These have many advantages that have be enumerated. Here we will look at the common and more recent transient techniques. One salient advantage of the transient technique is the storing of the information contained in the functional form of the time response. Analysis of the time domain function often yields information about the basic mechanisms. Here, we will analyze the information contained in the analysis of photoconductive decay. This chapter will focus on two techniques that have yielded much information about recombination mechanisms involved in the recombination process.

6.1.1. *Microwave Photoconductive Decay*

Microwave reflection is widely used to measure the recombination lifetime in semiconductors. This section will describe both the theoretical basis of the technique and present some typical experimental data. The technique is limited to materials that fall within a limited range of conductivity. The technique was first introduced by Deb and Nag[1] in 1962. The technique was further developed by Kunst and Beck[2] who developed a complete theory for semiconductor lifetime measurements. We will review the theory of microwave reflection photoconductive decay technique (μPCD).

This is a popular technique that is relatively easy to setup in a laboratory. It was introduced into the NREL laboratories as a standard method to measure carrier lifetime in silicon and other popular materials. Also, μPCD has undergone extensive comparison with other techniques at the NREL laboratories.

This technique been commercialized in a lifetime scanning and mapping apparatus that was designed primarily for silicon wafers. An advantage of the technique is the rapidity of measurement that has resulted in production of a commercial lifetime-mapping apparatus. One limitation of the technique is that it is limited to low-injection measurements. This inherent behavior results from microwave reflection being a nonlinear function of sample conductivity.

The μPCD measurement is basically a pump-probe technique with a pulsed generator of electron-hole pairs combined by a probe using the reflection of microwaves in 5 to 20 GHz frequency range. The transient microwave reflection is monitored in real time after the excitation of photo-carriers by the pulsed light source. The latter is often a pulsed laser as very short excitation pulses can provide the ability to measure lifetimes down to the nanosecond time domain when combined with a high- speed microwave detector. Here, we will analyze the system behavior and calculate the simulated transient response over a range of injection levels and background conductivity levels.

Figure 6.1 shows a schematic diagram of the NREL apparatus. A microwave generator, operating at either 7, 10 or 20 GHz, is used as a probe for photoconductive decay after excitation by a variety of nanosecond pulsed light sources.

The probe-beam reflection coefficient is a function of the conductivity of free carriers in the test sample, both the background doping and photocarriers.

The microwave generator uses a miniature Gunn oscillator that operates at discrete frequencies of 7, 10, and 20 GHz with a typical power of 20 mW. A quartz plate, with a coating of transparent conducting oxide, terminates the waveguide at the input but allows optical pulses to excite photo-carriers in the sample. This TCO is conducting at 10 GHz but transparent to the visible or the near IR

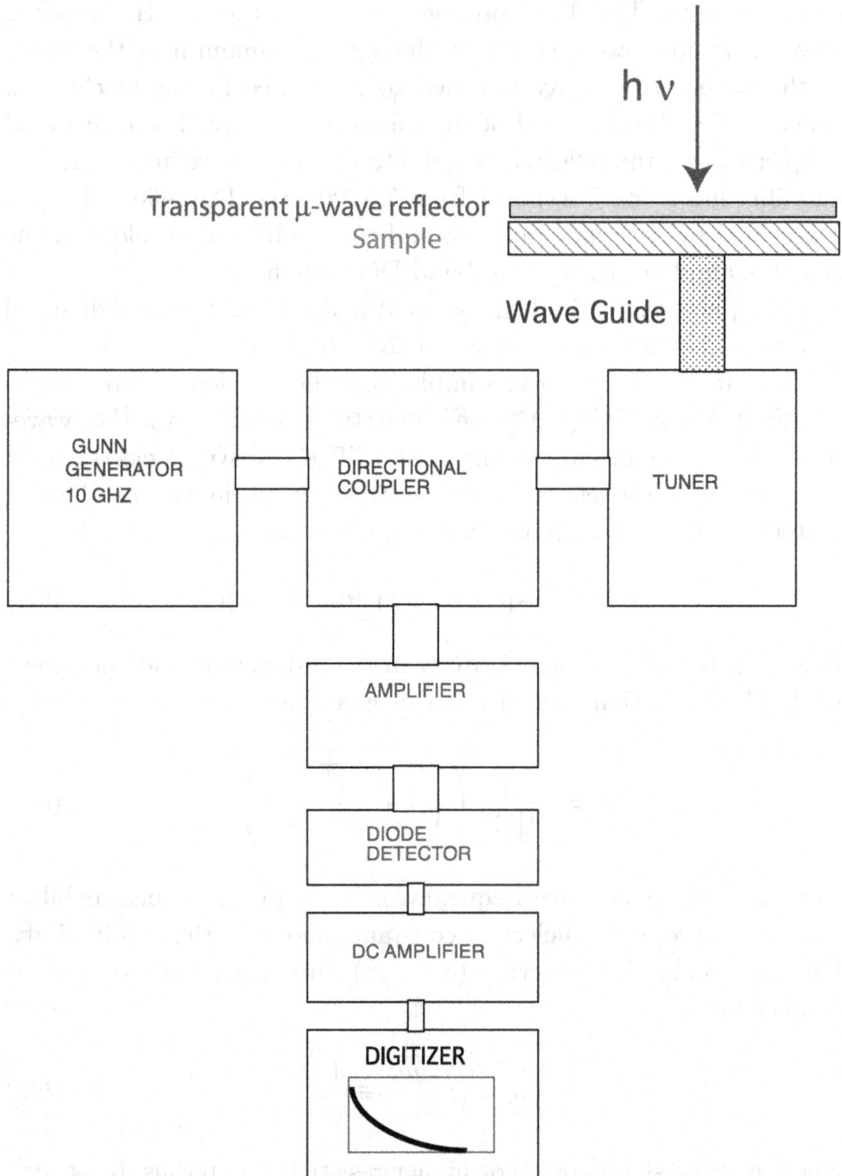

Fig. 6.1. A block diagram of the microwave PCD apparatus used by the author and colleagues for the data shown below. Three Gunn oscillator frequencies were used: 7 GHz, 10 GHz, and 20 GHz.

spectral range. The TCO produces a null in the 10 GHz standing wave in the microwave cavity producing a maximum near the center of the latter. The cavity is tuned to resonance by the mechanical tuner at the alternate end of the microwave cavity. The directional coupler directs the reflected signal into the microwave amplifier. The amplifier has a "flat" response from 5 to 20 GHz. The reflected signal is rectified by the diode detector and the modulated envelope of the signal is amplifier by a broad band DC amplifier.

The photo-signal is then averaged and digitized on a high speed transient digitizer, and transferred to a computer.

The interaction of the sample and the incident microwave is analyzed by applying Maxwell equations relative to the wave-propagation coefficients in the media. The transverse electric field component of the electromagnetic (EM) wave in the conducting material comes from the Maxwell equations as[3];

$$E = E_0 \exp(-\alpha x) \exp\left[i(\omega t - \beta x)\right]. \tag{6.1}$$

Here E is the transverse electric-vector component of the microwave field. The extinction coefficient, α, is given by:

$$\alpha = \omega \sqrt{\frac{\mu\varepsilon}{2}\left(\sqrt{1 + \frac{\sigma^2}{\omega^2\varepsilon^2}} - 1\right)}. \tag{6.2}$$

Here, ω is the microwave frequency, μ is the magnetic permeability, ε is the appropriate dielectric constant, and σ is the conductivity. For very conductive materials ($\sigma \gg \omega\varepsilon$), the expression reduces to a simpler form:

$$\alpha \approx \sqrt{\frac{\omega\mu\sigma}{2}} \equiv \frac{1}{\delta}. \tag{6.3}$$

where δ is the skin depth. For higher resistivity materials, ($\sigma \ll \omega\varepsilon$), one can approximate the extinction coefficient as:

$$\alpha \approx \frac{\sigma}{2}\sqrt{\frac{\mu}{\varepsilon}}. \tag{6.4}$$

For higher resistivity materials, such as semiconductors, the extinction coefficient scales with the absorption coefficient of the material.

The skin depth for any semiconductor as a function of sample conductivity is shown in Fig. 6.2. The measurement probe frequencies are A: 500 MHz, B: 10 GHz, and C: 20 GHz. The curves B and C represent microwave systems that exist in one of the laboratories of the PI. Curve A represents the common operating frequency of the RCPCD systems that operate in our laboratories and will be discussed in the next section. The 500 MHz shows the much larger penetration depth that is gained using a lower operating frequency. For thin film devices, the microwave probe easily penetrates the typical film thickness of several microns.

The discontinuity of optical constants at the air/semiconductor interface produces a partial reflection of the incident wave at the interface. The reflection and transmission coefficients at each interface are found from the continuity of the E and H components across the interface. The latter are perpendicular to the propagation vector and to each other. At the front surface, the reflection coefficient of

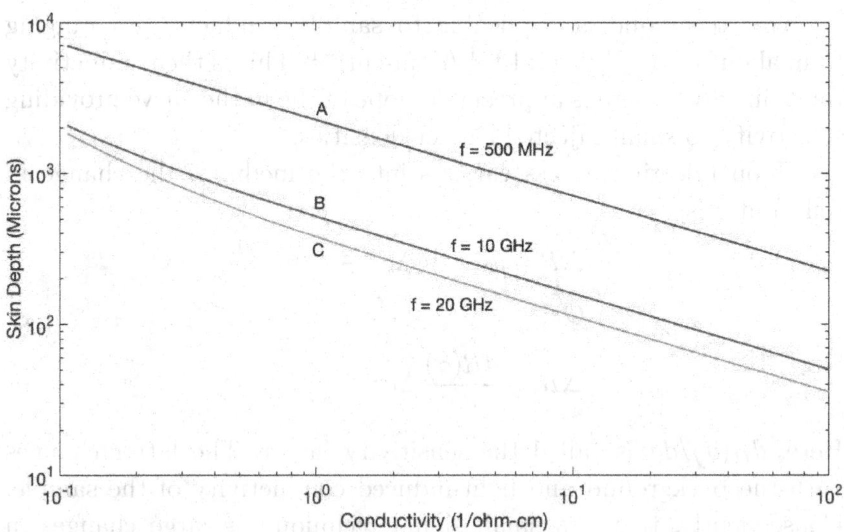

Fig. 6.2. The calculated skin depth for three probe frequencies. A. 500 MHz (used for RCPCD); B; 10 GHz; C. 20 Ghz.

the electric-field vector is found to be;

$$R = \frac{E_r}{E_i} = \frac{\eta_s - \eta_0}{\eta_s + \eta_0}. \tag{6.5}$$

Here, E_r and E_i are the reflected and incident electric-field components, respectively, of the EM wave. Also η_s and η_0 are the characteristic impedances of the semiconductor and vacuum, respectively. The impedances can be derived from the Maxwell equations as a function of frequency and conductivity:

$$\eta_s = \sqrt{\frac{i\omega\mu}{\sigma + i\omega\varepsilon}}. \tag{6.6}$$

For free space, η_0 is 377 ohms. The reflected power is then:

$$\frac{P_r}{P_i} = RR^*. \tag{6.7}$$

Figure 6.3 is a plot of the calculated reflected power as a function of conductivity with three incident microwave frequencies: A: 5 GHz; B: 10 GHz; C: 20 GHz. Here the static dielectric constant is taken as $\varepsilon = 12$ in the calculation.

This technique is applicable to sample conductivities ranging from about 5×10^{-4} to 1×10^{-1} (ohm-cm)$^{-1}$. This is the conductivity range in which there is appreciable slope of the in the curve providing sensitivity to small injected carrier densities.

Upon injecting excess carriers into the medium, the change in reflectance is:

$$\begin{aligned} &\Delta R = R(\sigma + \Delta\sigma) - R(\sigma); \\ ⩔ \\ &\Delta R = \frac{dR(\sigma)}{d\sigma}\Delta\sigma; \end{aligned} \tag{6.8}$$

Here, $dR(\sigma)/d\sigma$ is called the sensitivity factor. The latter changes with the background and light-induced conductivity of the sample. Consequently, this is a small signal technique as large changes in conductivity will not be linearly followed in reflectance changes. The minimum time response of the system is a function of the convolution

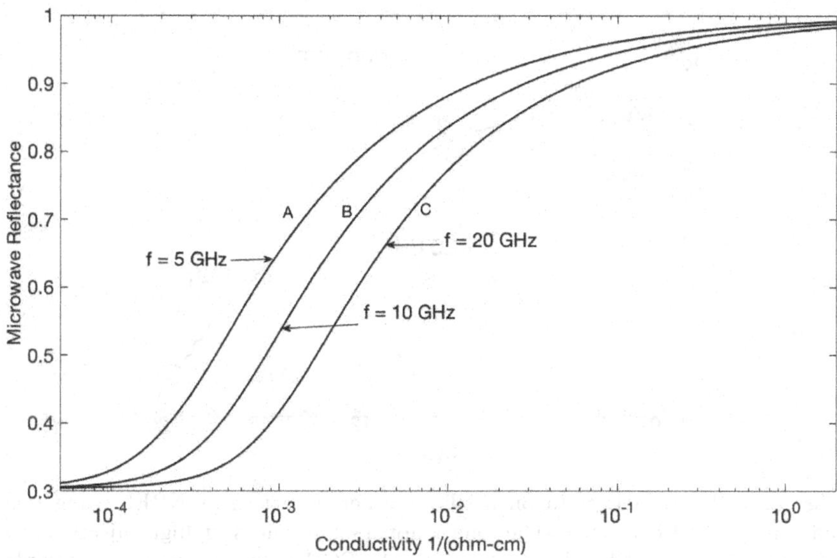

Fig. 6.3. The calculated microwave reflection coefficient versus material conductivity.

of the laser pulse width and the response of the amplifier electronics. One can achieve response times of several nanoseconds with the appropriate high frequency amplifiers.

6.1.2. *Nonlinear Effects of High Injection*

Figure 6.4 shows the nonlinear effects on the transient microwave reflection response at high carrier injection. The sample here was an unpassivated commercial silicon wafer with a p-type dark conductivity of 0.34 ohm cm^{-1}. This corresponds to a *p*-type doping level of $5 \times 10^{15}\ \text{cm}^{-1}$. An optical injection pulse was applied using 1100 nm wavelength pulse. The long and weakly absorbed wavelength assured uniform volume injection of carriers. The initial injection level was calculated as $5 \times 10^{16}\ \text{cm}^{-3}$ and therefor the injection was about ten times the background doping. The decay was measured in the NREL system with probe frequencies of A: 20 GHz and B: 7 GHz. The slope of the lifetime indicates a high-injection lifetime of about 15 μs at 20 GHz and 23 μs at 7 GHz. The low-injection lifetime is about 2 μs in both cases. These effects at high injection could be related to the

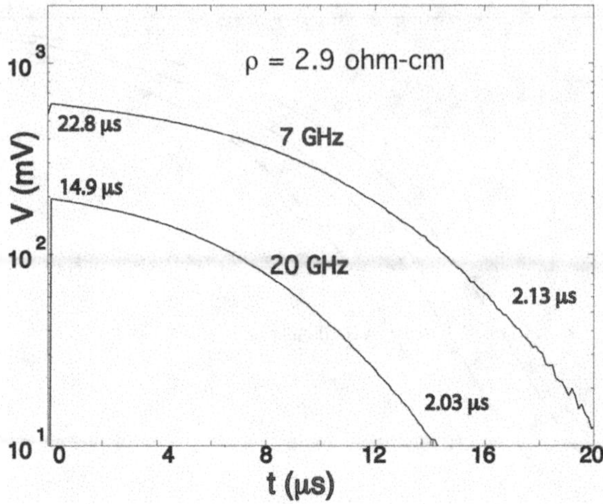

Fig. 6.4. The μPCD data on a silicon wafer measured at NREL using two different probe frequencies. The longer apparent lifetimes at high injection are an artifact produced by the nonlinearity of μPCD over wide ranges of sample conductivity.

saturation of deep level impurities as described in an earlier chapter. However, the size of that effect should not vary with probe frequency, indicating that is a measurement artifact. The same sample was measured with the same light pulse using the RCPCD technique that is described below. That technique was used to measure the sample using the same light pulse, and the measured high injection decay time was about 4 μs. The RCPCD technique will be shown to be linear over a wide signal range, so the large injection lifetime is an artifact of the nonlinearity of microwave reflection and the resulting small dynamic range. The probe frequency effect can be seen by comparing the curvature of the of the reflectance-conductivity curve, Fig. 6.3, at the operating conductivity change. Consequently, the μPCD technique must be limited to low injection conditions.

Transient decay is the primary method for measuring lifetime in germanium wafers as the indirect bandgap does not easily allow time-resolved photoluminescence to be measured. Germanium wafers are important in space photovoltaics as it is the dominant substrate used for the ultra-high efficiency tandem cell. In recent years, an active

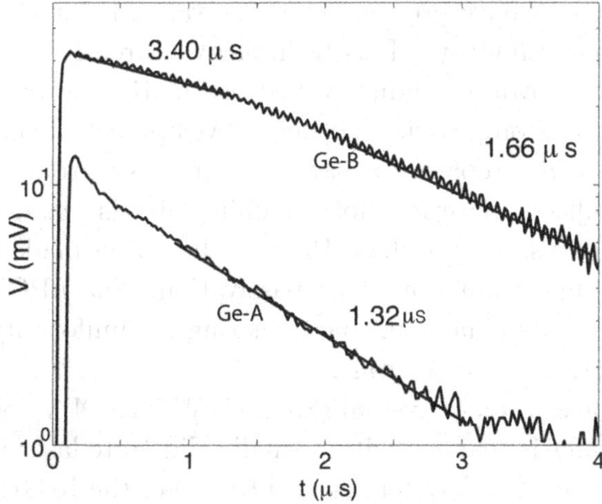

Fig. 6.5. The μPCD data measured on two heavily doped germanium wafers using a probe frequency of 10 GHz and an excitation wavelength of 1800 nm.

pn junction is grown into the Ge-substrate. This junction adds to the cell voltage and current, and therefore the substrate lifetime is important.

Microwave reflection was one of the first methods to measure carrier lifetime in Ge-wafers.[4] The Ge-wafer is the preferred substrate for the very high efficiency III-V tandem cell used in space photovoltaics. Some of the current cells use a fabricated junction in the wafer as the lowest bandgap cell in the tandem structure. Therefore, the lifetimes in the wafer are important in device performance.

Figure 6.5 shows μPCD data on two heavily doped germanium wafers measured at NREL. The doping level is in the $1.0 \times e^{18}\,\mathrm{cm}^{-3}$ range and are typical of those used as substrates for tandem cells. The sensitivity factor is low in this conductivity range and therefore the signals are relatively weak. Auger recombination is a dominant factor in this doping range.

6.2. μPCD Mapping

The data acquisition time for μPCD is quite rapid and has enabled the development of commercial lifetime mapping systems. These mapping systems have proven to be very valuable for the multicrystalline silicon technology. This technology currently dominates the residential photovoltaic industry because of the low-cost and high quality (relative long carrier lifetime) development of the materials. These wafers are prepared by sawing thin wafers from large boules of die-cast silicon material. Spatial uniformity is an inherent issue in these polycrystalline wafers. Progress has been made in low-cost purification and grain boundary passivation. The μPCD mapping systems are quite valuable in characterizing the uniformity and grain boundary activity of these wafers.

In the measurement system (Semilab WT-2000PV for the study here), the wafer is scanned with a small solid-state laser (904 nm) as the light source. A coil system is used to collect the 10 GHz reflection from each illuminated area. An algorithm is used to extract the lifetime from each measurement and a map is composed from the stored data. The measurements shown below were made in 2010 at NREL and were typical of material quality at that time. The sample shown is 5×5 inches in lateral dimension and about 300-μm-thick. This wafer has not undergone any post-growth treatment.

The lifetime values are color-coded with the legend shown at the right. The majority of the regions show a lifetime of 1.5 to 2 μs, but a defective area is observed at the left side of the wafer, apparently from some external contamination. The lifetimes in this region are about 0.2 to 0.4 μs. This system is designed for silicon but can be used for other materials that can be photo-excited by the 904 nm light source.

Again, the rapid data acquisition time allows the lifetime mapping application of microwave photoconductive decay. Because of the reflection required, a planar surface is required for μPCD and the reflection signal is not efficiency collected from irregular and non-planar sample.

6.2.1. *Transient μPCD Measurements with DC Light Bias*

Schmidt[5,6] and other authors have discussed the use of DC light bias on transient microwave decay measurements. A primary reason for

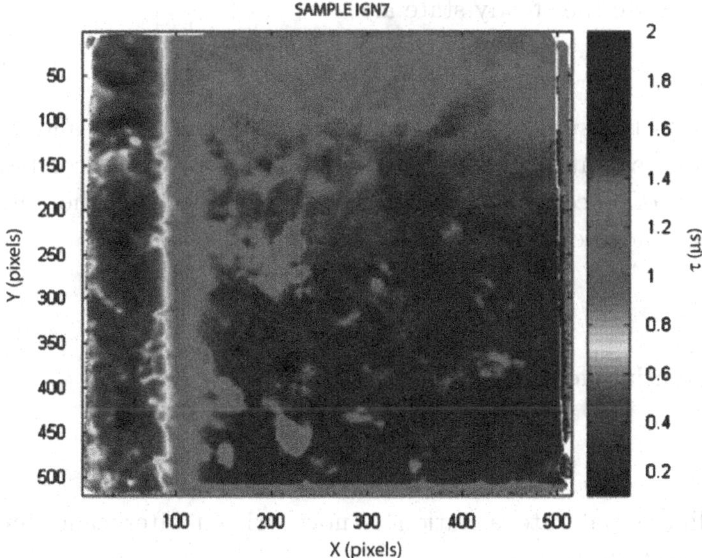

Fig. 6.6. The lifetime map of *s* multicrystalline silicon wafer using the Semilab scanner based on a μPCD sensor.

light bias is that the DC bias light will fill shallow trapping states so that only trapping by deep recombination sites are seen during the pulsed measure. This procedure is sometimes given the designation "trap filling". Another reason is related to the low dynamic range of the microwave reflection method. One can produce a high injection situation and then superimpose a weaker pulse to study the high injection decay dynamics.

We will use a procedure similar to that of Schmidt to analyze this superposition of excitation sources. When we look at Eq. (6.10) and assume that a DC light bias is superimposed on the pulsed laser excitation in a lifetime experiment, we write:

$$\frac{\partial \Delta n}{\partial t} = G - R = G_{ss} + \Delta G_p \delta(t) - R\left(n_{ss} + \Delta n(t)\right); \tag{6.9}$$

Also:

$$R(n_{ss}) = \frac{n_{ss}}{\tau(n_{ss})}; \tag{6.10}$$

$$n(t) = n_{ss} + \Delta n(t); \quad \Delta n(t) \ll n_{ss}; \tag{6.11}$$

We can write the steady state as:

$$n_{ss} = G_{ss}\tau(n_{ss}); \tag{6.12}$$

Here G is the sum of G_{ss} which is the DC bias light, and ΔG_p is the short excitation pulse that occurs at $t = 0$. The recombination function of excess carrier density is expanded by the binomial theorem and the first two terms are retained.

$$R(n_{ss} + \Delta n(t)) \approx R(n_{ss}) + \left[\frac{\partial R(n)}{\partial \Delta n}\right] \Delta n(t); \tag{6.13}$$

A differential lifetime can be defined:

$$\tau_d \equiv \left[\frac{\partial R(n)}{\partial \Delta n}\right]^{-1}; \tag{6.14}$$

The differential rate equation is now written after cancelling the steady state terms:

$$\frac{\partial \Delta n}{\partial t} = -\left[\frac{\partial R(n)}{\partial \Delta n}\right]_{n_{ss}}^{-1} \Delta n(t) = -\frac{\Delta n(t)}{\tau_d}; \tag{6.15}$$

$$\Delta n(t) = \Delta n_0 \exp(-t/\tau_d); \tag{6.16}$$

The measured transient is the differential lifetime defined as above in Eq. (6.17).

As an example, we can look at the case of a high-resisitivity material such as undoped, float zone silicon. It will be shown later here that the Auger recombination rate under high optical injection is:

$$\begin{aligned} &R_A \approx C_n (n_{ss} + \Delta n)(p_{ss} + \Delta p)^2; \\ &where: \ n_{ss} = p_{ss}; \Delta n = \Delta p. \\ &Or: \ R_A = C_n (n_{ss} + \Delta n)^3. \end{aligned} \tag{6.17}$$

The electron Auger lifetime under light bias is:

$$\tau_A = 1/C_n n_{ss}^2; \tag{6.18}$$

This neglects the small contribution from the pulsed source. By applying Eq. (6.15), the differential lifetime is one third of the real lifetime in this case:

$$\tau_d = 1/3C_n\,(n_{ss} + \Delta n;)^2 \approx \frac{\tau_A}{3}; \tag{6.19}$$

This effect was observed experimentally by Schmidt in measuring high resistivity silicon using the light bias technique and microwave reflection. Those data showed the differential lifetimes were 1/3 of those measured by QSSPC.

6.3. Resonant Coupled Photoconductive Decay (RCPCD)

The RCPCD technique was invented[7–9] at NREL in the late 1990s. This method was developed to measure the transient recombination lifetime in indirect bandgap semiconductors, such as silicon, to complement the TRPL facility.[10] However, the technique has been effectively applied to a wide range of direct and indirect bandgap materials.[11]

RCPCD has the advantages of being contactless and linear over a large dynamic range. This allows measurement over a wide injection range without light bias. RCPCD is also very sensitive to small and irregular samples. In our laboratory, we have found it to be more sensitive than μPCD. The latter attribute is especially useful for small volume materials such as thin films. The technique is easily applicable to any sample size or shape and does not require planar samples. It was invented in the 1990s to measure carrier lifetime in small (~1 mm diameter) silicon spheres that were being fabricated into photovoltaic cells at that time.

Another advantage is the application to small bandgap materials of any band structure. The direct small bandgap semiconductors are not easily measureable because of the lack of sensitive infrared photo detectors that allow single photon detection. Materials such as HgCdTe with bandgaps in the 0.1 eV range have been analyzed by the technique.

A block diagram of the RCPCD apparatus is shown in Fig. 6.7(a). The sample is placed on the vertical positioner. The central feature of the system is a platform that can continuously vary the coupling-coefficient to the coil antenna that provides a radio frequency signal.

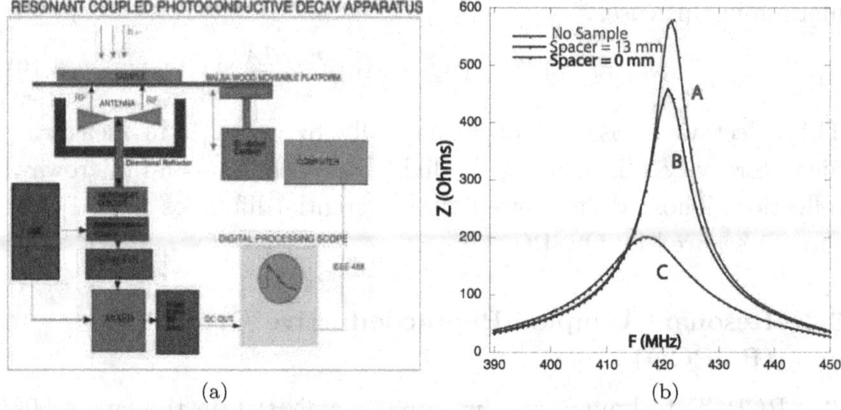

Fig. 6.7. (a) RCPCD Block Diagram. (b) Impedance-frequency response of silicon wafer of about 100 ohm-cm resistivity. [Reprinted with permission from Ahrenkiel, R. K., *J. Vac. Sci. Technol.* **B31**, pp. 04D113-1 to 04D113-8. Copyright [2019], American Vacuum Society.]

The central frequency used in the initial setup is 425 MHz but frequencies up to 1000 MHz have been employed. This frequency range has appreciable penetration depth in bulk wafers of silicon and other common materials of typical doping.

The platform of the positioner is made is thin balsa wood that has a very low loss at the frequency range used in these measurements. This positioner is adjusted to produce the same impedance-frequency (Z-F) curve for a wide range of samples sizes and conductivities. The operating frequency is then adjusted so that the point of largest slope of the Z-F curve is used for the measurement. Figure 6.7(b) shows the Z-F curve as measured by an impedance analyzer with no sample on the sample platform (A) and with a 100 ohm-cm on the moveable platform (B and C). As the distance between the sample and coil are decreased, the resonance frequency and Q decrease. For a given sample, the platform spacing is changed to give the optimum response curve. As the Q decreases, the sensitivity decreases but the response time becomes faster. Therefore, a compromise between sensitivity and response time can be achieved by means of the mechanical positioner.

Figure 6.8(a) is an equivalent circuit representation of the drive circuit and the inductively coupled sample that is used to fit the

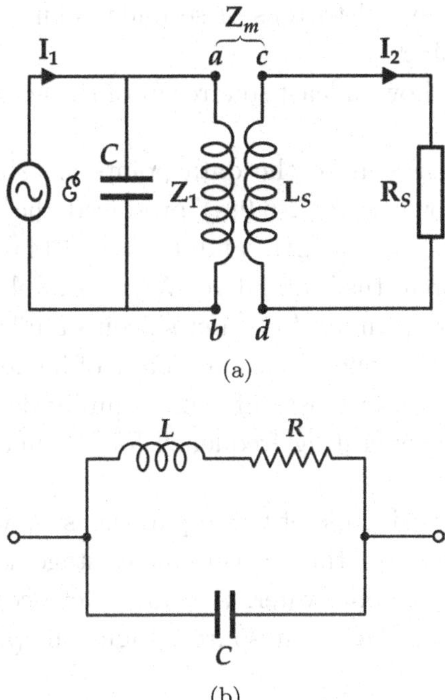

Fig. 6.8. (a) The equivalent representation of the drive circuit and the inductively coupled sample that is used to fit the response of the actual system. (b) The reduction of circuit of 6.8(a) to a simple RLC circuit for modeling purposes. [Reprinted with permission from Ahrenkiel, R. K., *J. Vac. Sci. Technol.* **B31**, pp. 04D113-1 to 04D113-8. Copyright [2019], American Vacuum Society.]

response of the actual system. The case of no sample is represented by R_s approaching infinity. There are small resistive losses in the circuit components, so there is some finite loss in the system. This circuit can be reduced to the simple RLC circuit of Fig. 6.8(b) where R_s is the effective resistance that is coupled into the circuit. Here R is related to the resistive loss of the sample that is placed on the positioner and has a very small effective value when there is no sample. By changing the spatial coupling, the excess carrier decay time can be measured over a wide range of injection level. In other words, the system a large dynamic range as the conductivity change can be brought into the linear response range by changing the spatial coupling. A disadvantage of the technique is that the data collection time is typically several to tens of seconds. Thus, it is not suitable for efficient mapping.

Figure 6.9(b) shows a least squares fit of the actual Z-F data with no sample.

The least squares fit to the data points are shown in the solid curve. The fit shown in Fig. 6.9(a) produced the following values: $R = 0.36$ ohms, $L = 5.44$ nH, and $C = 26.14$ pF. The intrinsic Q-factor of the apparatus is about 40. With a sample in place, we find a Q of about 16 is optimum for wafer silicon samples. The operating frequency is chosen as one of the two points of highest slope or about 424 MHz. Very large decreases of signal amplitude are observed for small variations of operating frequency for the maximum inflection point.

The operating principle of the apparatus is shown in Fig. 6.9(b). The sample placed on the measurement stage is a high-quality, high-resistivity float zone wafer. A small microscope illuminator is used as a low power DC light source. Figure 6.9(a) shows the F-Z

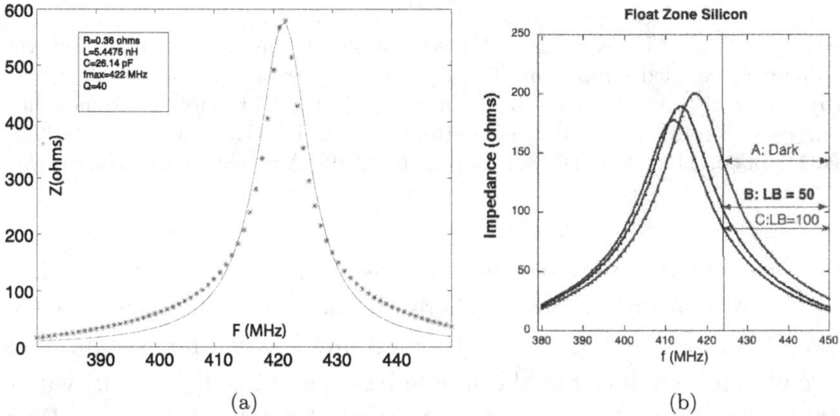

Fig. 6.9. (a) The measurement of the system response of the system with a high frequency impedance meter. The data is shown a points on the curve. A computer fit to the data is the solid curve using the circuit of Fig. 6.8(b). (b) The system response with a float zone silicon wafer on the stage. Curve A is in the dark; Curve B: Impedance with white light illumination at one half power. Curve C impedance with white light at full power. [Reprinted with permission from Ahrenkiel, R. K., *J. Vac. Sci. Technol.* **B31**, pp. 04D113-1 to 04D113-8. Copyright [2019], American Vacuum Society.]

response with the sample in the dark. Figure 6.9(b), curves B and C show the F-Z response with the DC source set as one-half (B) and full intensity (C). By monitoring the sensor at 424 MHz, one sees the impedance drop by about 40 ohms at 1/2 bias and 70 ohms at full bias. The apparatus is electrically connected to an impedance bridge, and the change in impedance transferred to a detection device such a transient digitizer. The sharp resonance produces the high sensitivity of the system. One can increase the sensitivity by using a higher-Q condition but there is a loss of high frequency response (risetime) to pulsed signals. One can calculate the minimum response time as:

$$\tau_r = \frac{Q}{\omega}. \quad \tau_r = \frac{Q}{\omega} \tag{6.20}$$

The risetime increase with increasing Q and decreases with frequency. The risetime is the metric used to calculate the minimum lifetime that can be measured. In practice, a compromise is reached that is appropriate to the material. For the Q of 16 used here and an operating frequency of 424 MHz, the response time is 37 ns. This response is considerably longer than that of the microwave PCD discussed previously, but the sensitivity to small signal is larger. A choice is made as to which system is most appropriate to a give measurement.

In these examples, Fig. 6.10(a) shows the RCPCD data of two, commercially available germanium wafers that are five inches in diameter. The sample positioner is raised until the coupling is reduced to provide the desired Q-factor that is used for the measurement. Figure 6.10(b) gives the data for a thin film GaAs DH that is several microns in thickness on an undoped GaAs substrate. In this case, the sample positioner if adjacent to the sensor as the sample size is very small.

6.4. Comparison with μPCD

In Fig. 6.4, the data were shown on the μPCD of a silicon wafer that had a resistivity of 2.9 ohm-cm. The sample was remeasured with the RCPCD setup using the same light source and pulse intensity. The

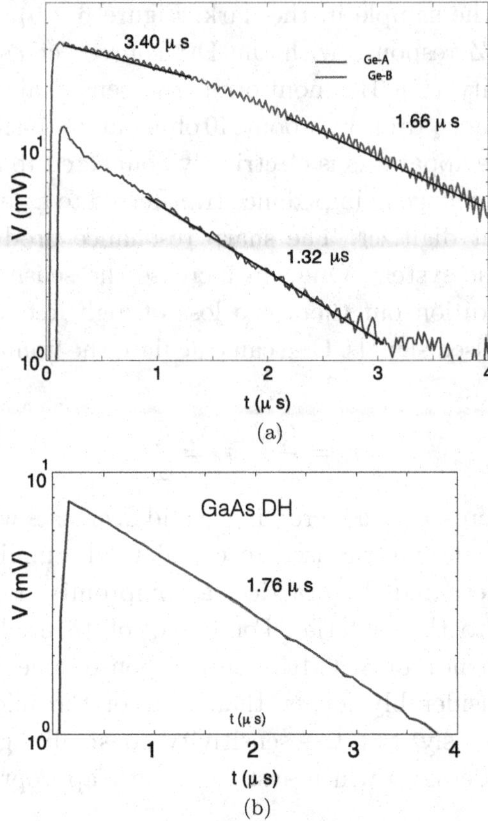

Fig. 6.10. (a) RCPCD measurement of two germanium wafers. (b) The RCPCD photoconductive decay of an undoped GaInP/GaAs double heterostructure at low injection.

data from the RCPCD and the 20 GHz μPCD are shown in Fig. 6.11. One sees that the low injection lifetimes are the same, about 2 μs, whereas the high-injection lifetime is about 4 μs for the RCPCD case. The larger lifetime at high injection is expected for silicon owing to the saturation of deep defects at either the bulk or surface. However, the μPCD data reflects the nonlinearity and decrease in sensitivity inherent in the technique at higher conductivity levels. Thus, the μPCD data in inaccurate as the high injection levels shown here. The μPCD technique must be limited to low-injection conditions such as those found in the mapping application. However,

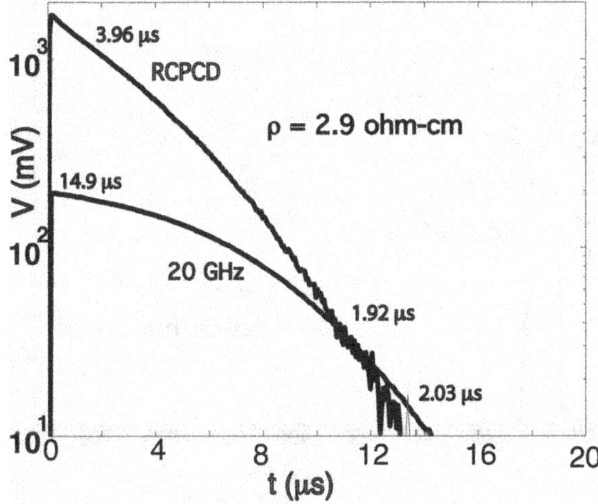

Fig. 6.11. The RCPCD data and the 20 GHz μPCD data of the sample measured in Fig. 6.4. The 20 GHz is distorted by nonlinearity at high injection.

data acquisition is much faster with μPCD than with RCPCD, and that attribute makes the mapping application viable.

6.5. Experimental Measurements of SRV

The RCPCD method is very valuable for the measurement of the SRV of samples immersed in a passivating liquid. In situ measurements are particularly easy as the sample is placed in an insulating beaker containing the sample and the entire system is placed on the sample stage. One example is the use of the iodine/methanol solution to passivate silicon.

Figure 6.12 shows the passivating effects of a silicon wafer immersed in a solution of iodine/methanol. The iodine chemically interacts with unbounded electrons on the surface atoms on the silicon wafer. These bonds are saturated and electronically passivated, thus eliminating surface states.[12] This is a temporary passivation process that allows the detection and characterization of bulk recombination centers. In a final device, the surface states are passivated by a permanent process such as the growth of a thermal oxide, such as SiO_2.

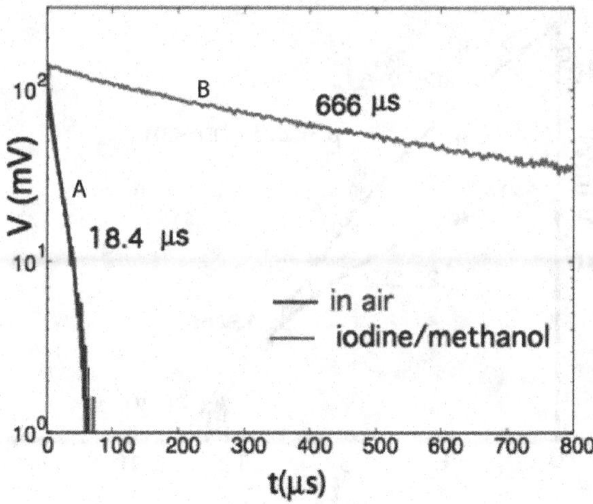

Fig. 6.12. Photoconductive decay data for a float zone silicon wafer. A: Measurement in air. B: Measurement made while wafer immersed in passivating solution of iodine/methanol. The latter reduces the surface state density by orders of magnitude.

For a given surface density N_{st}, the surface potential V_s will depend on the doping density N and the injection level ρ/N. Therefore, S will be dependent upon the doping density and injection level. Experimentally, S is usually found to be doping dependent. In addition, S depends on the injected carrier density ρ because surface states may be temporarily "filled" by excess carriers. The dependence of surface density ρ_s on the injection level is described by Dhariwal and Mehrotra.[13] This effect is produced by the injection level dependence of the surface SRH recombination.

The surface may be passivated electrostatically by intentional band bending that forces minority carriers away from the surface.[14,15] One such example is the significant band bending at the surface by depositing SiNx:H on the surface of a silicon wafer. SiN_x:H is used as a passivating insulator and anti-reflection coating on silicon photovoltaic devices. Here, SiNx:H was deposited on a high resistivity float zone (FZ) wafer. The background doping of this particular wafer is n-type in the range of $6 \times 10^{13}\,cm^{-3}$. It is typical for deposited SiN_x:H to contain a large density of fixed positive charge.

Capacitance voltage (C-V) studies of these devices showed a "flat band" voltage shift of about −9 volts, indicating an accumulation of electrons at the semiconductor-insulator surface. This accumulation layer then assures that there is a large negative band bending at the interface. The negative band bending repels minority holes from the interface and reduces recombination at interface states. RCPCD measurements were made on this wafer with an incident wavelength of 532 nm. The pulsed injection produces an initial injection level of 4.75×10^{15} electron-hole pairs per cm^3. The latter is the calculated electron-hole density after the initial distribution becomes "flat" throughout the volume because of diffusion (~20 microseconds). Curve A shows the photoconductive decay (PCD) with no white light bias applied. The decay time is about 6.8 ms, indicative of low defect density in the bulk of the sample as the minority holes are separated from the majority electrons. In curve B, 106 mW/cm^2 white light bias is superimposed on the pulsed laser signal. In this case, the light bias intensity is weak as compared to the pulse laser intensity. Or: $G_{ss} < G_p$. The primary effect of the light bias is to generate

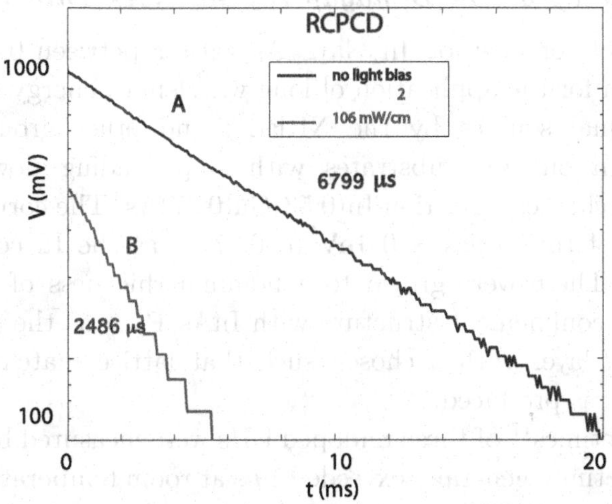

Fig. 6.13. The photoconductive decay of a silicon wafer that had a layer of SiN deposited on each surface. The data measured in the "dark" and with a white light bias.

electron-hole pairs that neutralize the band banding from the fixed charge in the SiN.

A dc signal level shift in the RCPCD detection system is about the same as the initial value for the pulsed data of curve A. Thus, the bias light intensity is approximately equal to the peak pulse intensity. The carrier lifetime decreases to about 2.5 μs. The white light bias "flattens" the band bending such that minority holes can interact with the interface states. The carrier lifetime decreases from 6.8 μs to 3.0 μs in this "flat band" situation. Here, the pulse intensity is larger or equal to the bias intensity, in contrast to the previous derivation. In summary, the "flat bands" reduces the effective SRV in accordance and reverses the electrostatic passivation.

These data contrast with the predicted results of the previous derivation of the effects of light bias, where the bias light intensity is much larger than the pulse intensity. In this case, the electrostatic band bending was a significant source of the isolation of the carriers from the surfaces. This contrast clarifies the source of the experimental observation. The source of the electrostatic band bending is the charged insulator that was deposited on the surfaces.

6.5.1. *Undoped Low Bandgap $In_xGa_{1-x}As$ Thin Films*

Materials of composition $In_xGa_{1-x}As$ with x between 0.68 to 0.80 were grown for the application of long wavelength energy generation from thermal sources by the NREL[16] and other groups. These were grown on InP substrates with step grading down to the lattice-matched composition In(0.53)Ga(0.47)As. The corresponding bandgaps of this series is 0.6 eV to 0.47 eV as the In content was increased. These were grown to a nominal thickness of 2 microns into a DH confinement structure with $InAs_yP_{1-y}$ as the window or passivating layer with y chosen such that lattice matching to the active layer is produced.

The lifetimes[17] of these undoped DHs were measured by RCPCD and the lifetimes generally exceeded 1 μs at room temperature. These data are shown in Fig. 6.14. Curve A shows data measured on a film with band gap of 0.57 eV. These data show a low-injection lifetime of 4.84 μs and a high-injection lifetime of 8.23 μs. The bandgaps on

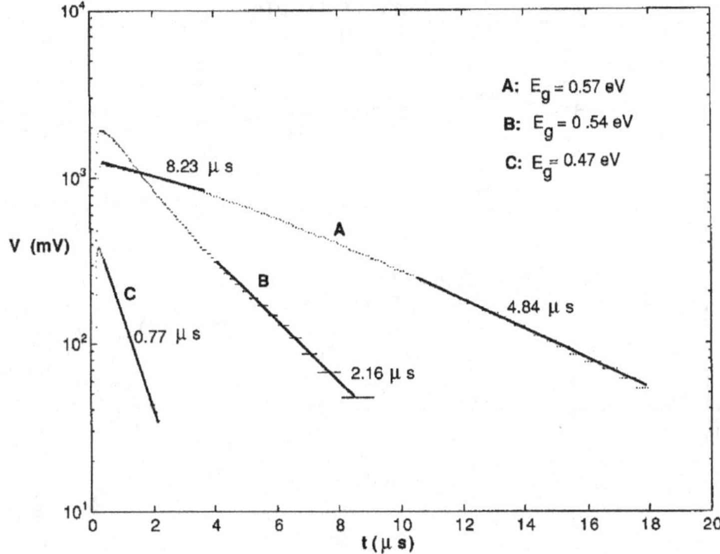

Fig. 6.14. RCPCD data on three epitaxial thin films of $In_xGa_{1-x}As$ with the bandgap indicated in the figure. These were step-graded to In(0.53)Ga(0.47)As which was grown on an InP substrate. [Reprinted from Ahrenkiel, R. K., Johnston, S. W., Webb, J. D., Gedvilas, L. M., Carapella, J. J., and Wanlass, M. W. (2001). *Appl. Phys. Lett.* **78**, 1092, with permission of AIP Publishing.]

sample B and C were 0.54 eV and 0.47 eV, respectively. The decay behavior is typical of a SRH-dominated recombination process.

6.5.2. *Polycrystalline Thin Films*

The photoconducting properties of a polycrystalline CdTe thin film is illustrated in Fig. 6.15. The sample is a thin film of CdTe that is deposited by the close-space sublimation process and is the base of a CdTe solar cell. The film is about four microns thick and is deposited directly on a glass substrate coated with a thin film of CdS. The CdTe is not intentionally doped but is lightly p-type owing to native acceptors.

Figure 6.15 shows the RCPCD signals that result from pulses from a tripled YAG laser (355 nm). Curves B results from the unattenuated pulse and Curve A from a decade of attenuation. Other studies confirm that the signal represents the signal from conducting

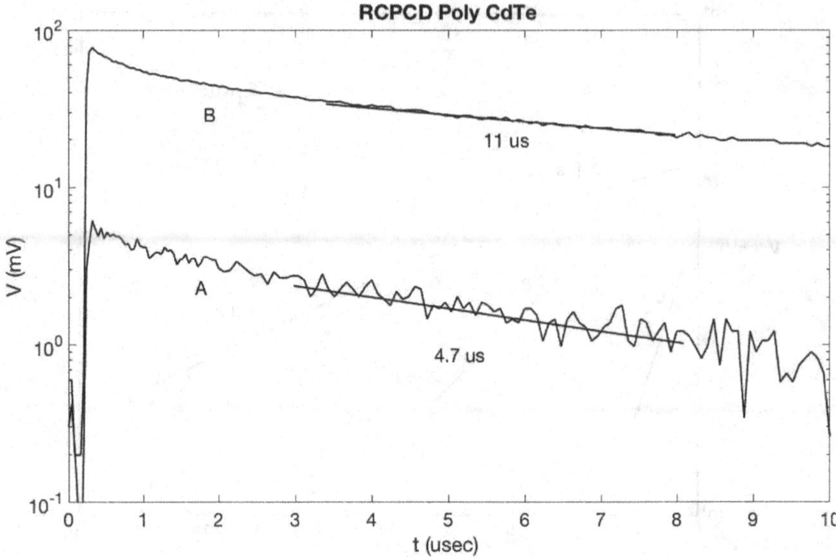

Fig. 6.15. The RCPCD signal from a thin film of polycrystalline CdTe that is in a solar cell configuration. The excitation wavelength is 355 nm. The signal is dominated by shallow trapping.

electrons that are being trapped and then thermally excited into the conduction band. The measured decay times represent the shallow trap regeneration rate from the traps. Measurements by time-resolved photoluminescence indicate that electron-hole recombination time for these films is one to two nanoseconds. The recombination process is "delayed" by the shallow trapping, but occurs on a time scale that is well below the time resolution of RCPCD. The shallow trapping regeneration rate becomes longer as the ambient temperature is lowered as will be shown in a later chapter. The RCPCD confirms the presence of shallow trapping but is unable to provide the recombination lifetime because of limited time response.

6.5.3. *Large Volume Samples*

The RCPCD technique is uniquely useful for large volume samples such as ingots and cast-silicon bricks. The lower probe frequencies in

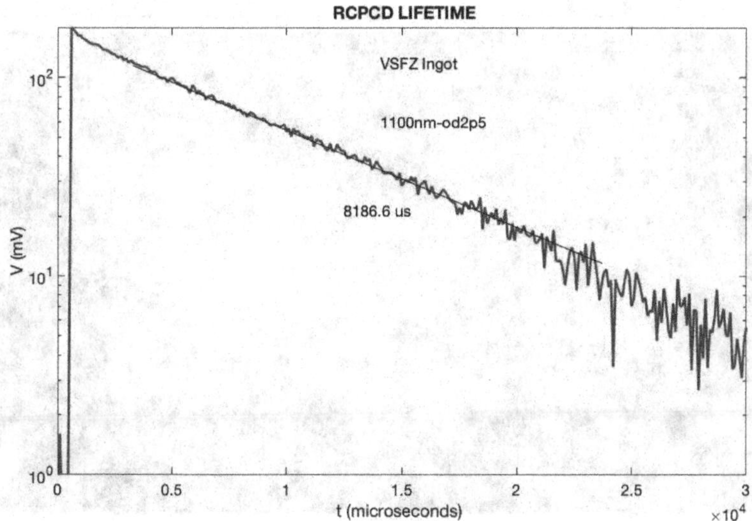

Fig. 6.16. RCPCD measurement of an undoped silicon ingot grown by the float zone technique. The excitation wavelength is 1100 nm.

the 400 MHz range can be combined with a weakly absorbed optical wavelengths to provide a probe that penetrates deeply into relatively large volumes.

An example is shown in Fig. 6.16 for an unpassivated, high-resistivity float zone ingot that is several cm in thickness. This measurement used the RCPCD technique at a probe frequency of 425 MHz and the skin depth is very large such that the rf sensor probes deeply into the ingot. The pulsed OPO light pulse is tuned to 1100 nm where the indirect, phonon-assisted transitions provide a very weak absorption. The reported absorption coefficient[18] is very small ($\alpha \sim 4.8\,\text{cm}^{-1}$) and the absorption is nearly uniform and deeply penetrating. The surface recombination component is negligible here and a relatively large lifetime of about 8 ms is found in the interior of the ingot.

S. W. Johnston[19] used a portable RCPCD sensor to map lifetimes in a cast-silicon "brick" that had dimensions of 156 mm × 156 mm × 265 mm in length. The OPO laser pulse was tuned to 1150 nm such that the absorption depth is several cm. A

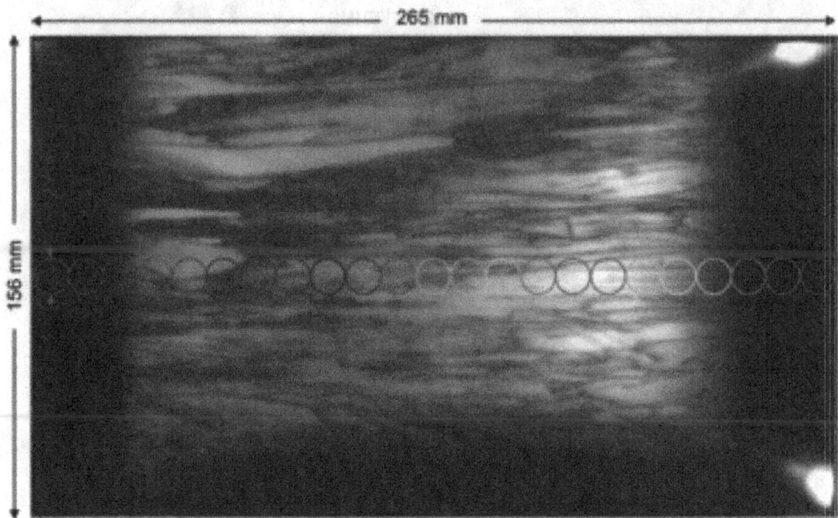

Fig. 6.17. PL image of silicon brick that was formed by casting in a crucible. The top is at the left and the bottom at the right. [© [2019] IEEE. Reprinted, with permission, from Johnston, S. W., Yan, F., Zaunbrecher, K., Al-Jassim, M., Sidelheir, O., and Ounadjela, K. (2010). 38th IEEE PVSC, 2010, pp. 406–410.]

photoluminescence image of the ingot is shown in Fig. 6.17. The image is captured with a Princeton Instruments Pixis 1024BR Si charge-coupled-device (CCD). The light source is an array of four 30-W, 810 nm laser diodes with diffusers to provide nearly uniform excitation density. The PL intensity is proportional to the doping concentration × lifetime product as will be discussed.

A small RCPCD sensor is scanned across the wafer with a pulsed OPO of several ns duration. The laser beam is passed through the center of the sensor coil. The electronic grade silicon brick is illuminated by the laser in the indicated spots. These spots show the location of lifetime measurements by the 12 mm diameter laser probe. The RCPCD apparatus is tuned to 420 MHz and the carrier decays at each location are plotted in Figs. 6.18(a) and 6.18(b). The measurement time is several seconds for each point. The OPO is tuned to 1150 nm so that the optical absorption is very weak[18] with $\alpha \sim 0.63\,\mathrm{cm}^{-1}$. The optical penetration depth $(1/\alpha)$ here is 16 mm and carriers are excited very deeply in the brick. The RCPCD

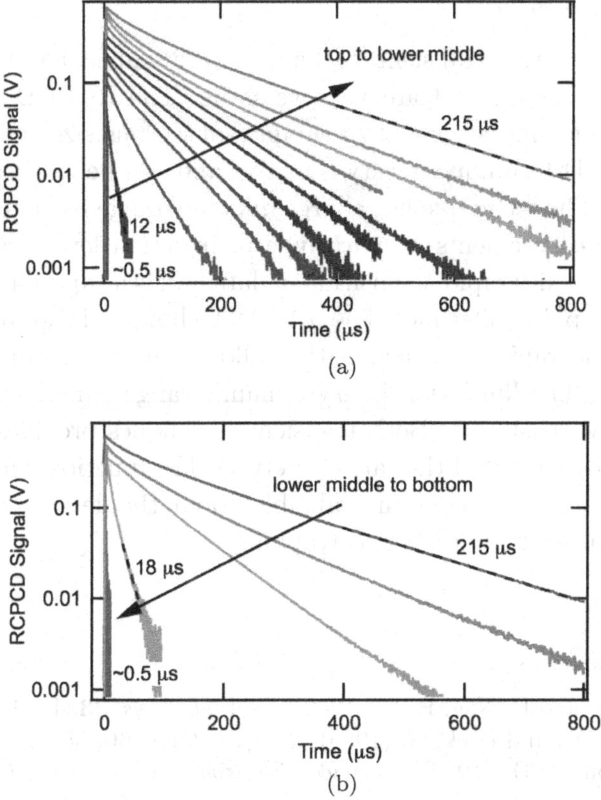

Fig. 6.18. (a) The RCPCD lifetimes measured in upper half of the brick shown in Fig. 6.16. The curves are correlated with the measurement points in Fig. 6.16. (b) RCPCD lifetime measurement measurement of the lower portion of the brick of Fig. 6.16. The measurement points are indicated in the figure. [© [2019] IEEE. Reprinted, with permission, from Johnston, S. W., Yan, F., Zaunbrecher, K., Al-Jassim, M., Sidelheir, O., and Ounadjela, K. (2010). 38th IEEE PVSC, 2010, pp. 406–410.]

transient is a direct measurement of the photoconductive decay that does not require any corrections to produce the indicated lifetime value.

The plots show that the lifetimes in the center of the brick are about 200 μs but degrade to about 1 μs at the top and bottom of the brick. This degradation is caused by impurities diffusing from the crucible and by the segregation of impurities to the surfaces.

6.5.4. *Summary*

In summary, the transient techniques are applicable to a wide range of materials and are very versatlie. The RCPCD technique is valuable for measurement of samples of various sizes and shapes. The RCPCD technique is very sensitive and can be applied to most materials. The technique has a large dynamic range by spatial tuning allowing measurements over a range of injection levels. Microwave (μPCD) provides rapid acquisition of data and is a superior technique for the mapping lifetime. The μPCD technique is quite easy to use and the rapid data acquisition allows lifetime mapping to be performed. The limitation is low dynamic range (small signal only) and limited sensitivity. Both transient techniques provides fast and direct measurement of the carrier lifetime. The functional form of the excess carrier decay contains valuable about the decay mechanisms involved in recombination and trapping.

References

1. Deb, S. and B. Nag, B. R. (1962). *J. Appl. Phys.* **33**, 1064.
2. Kunst, M. and Beck, G. (1986). *J. Appl. Phys.* **60**, 3558.
3. Jackson, J. D. (1999). *Classical Electrodynamics, 3rd Edition*, John Wiley and Sons.
4. Ramsa, A. F., Jacobs, H., and Brand, F. A. (1949). *J. Appl. Phys.* **30**, 1054.
5. Schmidt, J. (1999). *IEEE Trans. On Electron Devices* **46**, 2018–2025.
6. Giesecke, J. A., Glunz, S. W., and Warta, W. (2013). *J. Appl. Phys.* **113**, 073706.
7. Ahrenkiel, R., US Patent 5,929,652, July 27, 1999.
8. Ahrenkiel, R. and Johnston, S., US Patent 6,275,060, August 14, 2001.
9. Johnston, S. and Ahrenkiel, R., US Patent 6,369,603, April 9, 2002.
10. Ahrenkiel, R. K. and Johnston, S. (1998). *Sol. Ener. Mat. & Sol. Cells* **55**, 59–73.
11. Ahrenkiel, R. K., *J. Vac. Sci. Technol.* **B31**, pp. 04D113-1 to 04D113-8.
12. M'Saad, H., Michel, J., Lappe, J. J., and Kimerling, L. C. (1994). *J. Electron. Mater.* **23**, 487.
13. Dhariwal, S. R. and Mehrotra, D. R. (1988). *Solid-St. Electron.* **31**, 1355.

14. Agostinelli, G., Delabie, A., Vitanov, P., Alexieva, A., Dekkers, H. F. W., De Wolf, S., and Beaucarne, G. (2006). *Solar Energy Materials and Solar Cells* **90**, 3438–3443.
15. Ahrenkiel, R. K., Feldman, A., George, M., and Chandra, H. (2009). *Proceedings of the 2009 34th IEEE Photovoltaic Specialists Conference (PVSC 2009)*, p. 001054-9.
16. Wanlass, M., Ward, J. S., Emery, K. A., Al-Jassim, M. M., Jones, K. M., and Coutts, T. J. (1996). *Sol. Energy Mater and Sol. Cells* **41/42**, 405.
17. Ahrenkiel, R. K., Johnston, S. W., Webb, J. D., Gedvilas, L. M., Carapella, J. J., and Wanlass, M. W. (2001). *Appl. Phys. Lett.* **78**, 1092.
18. Edwards, D. (1985). in "Handbook of Optical Constants in Solids", p. 566, edited by E. D. Palik.
19. Johnston, S. W., Yan, F., Zaunbrecher, K., Al-Jassim, M., Sidelheir, O., and Ounadjela, K. (2010). 38th IEEE PVSC, 2010, pp. 406–410.

CHAPTER 7

Time-Resolved Photoluminescence: Techniques and Analysis

Radiative processes are the dominant component of recombination in direct bandgap semiconductors. These include most of the popular direct gap materials such as GaAs and CdTe. A standard approach for measuring the recombination parameters in GaAs and other direct materials is to generate electron-hole pairs with a very short laser pulse (nanosecond to femtosecond duration). The current approach to measuring the recombination parameters in GaAs and other direct materials is to generate electron-hole pairs with a very short laser pulse (nanosecond to femtosecond duration). Time-resolved photoluminescence (TRPL) has become the most widely used techniques for characterization carrier lifetime in direct bandgap semiconductors.

As TRPL becomes more popular and perhaps the most common technique across the vast spectrum of materials, the correct interpretation of decay curves has become increasingly important. The data interpretation issue was addressed by the first author in a 1993 review.[1] Recent publications by Maiberg[2] and Scheer have done a comprehensive analysis of TRPL carrier decay using numerical analysis of Poisson and drift-diffusion equations to analyze a variety of physical effects. The authors emphasized the shape of the curves on injection level, photon recycling, charge separation, deep traps, space charge and other common influences on carrier decay. They analyzed both decay dynamics from the steady state and short pulse excitation.

It was shown earlier, that the low-injection intensity of transient photoluminescence is:

$$\Delta I_{\mathrm{PL}}(t) = B * \Delta n_0 \exp(-t/\tau);$$

$$\textit{where:}$$

$$\frac{1}{\tau} = \frac{1}{\tau_{\mathrm{R}}} + \frac{1}{\tau_{\mathrm{NR}}} = B * N + \frac{1}{\tau_{\mathrm{NR}}}. \tag{7.1}$$

where B is the radiative coefficient that is specific to a material.

There are several techniques for doing this measurement. The analog method was first used and may still find some applications. Dual or multi-sensor techniques using TRPL will be described in a later chapter. However, rime-resolved single photon counting has become the preferred technique because of superior sensitivity and time resolution as shown in Fig. 7.1.

The basic method of measuring TRPL uses the simple analog system shown below.

This method is simple to use but lacks the sensitivity of photon counting. However, it is useful for some of the simultaneous methods in which TRPL and another parameter like photoconductivity are measured at the same time. These methods will be described in a later chapter. The pulsed laser usually has a repetition rate that is usually in the range from 10 to 10^3 Hz. The higher repetition rates allow better signal averaging. The collection optics must be sufficiently efficient to provide an adequate signal at the photo detector. This configuration is compatible with simultaneous measurement of photoconductivity or free-carrier absorption transients. An application of this system will be shown later in a later chapter.

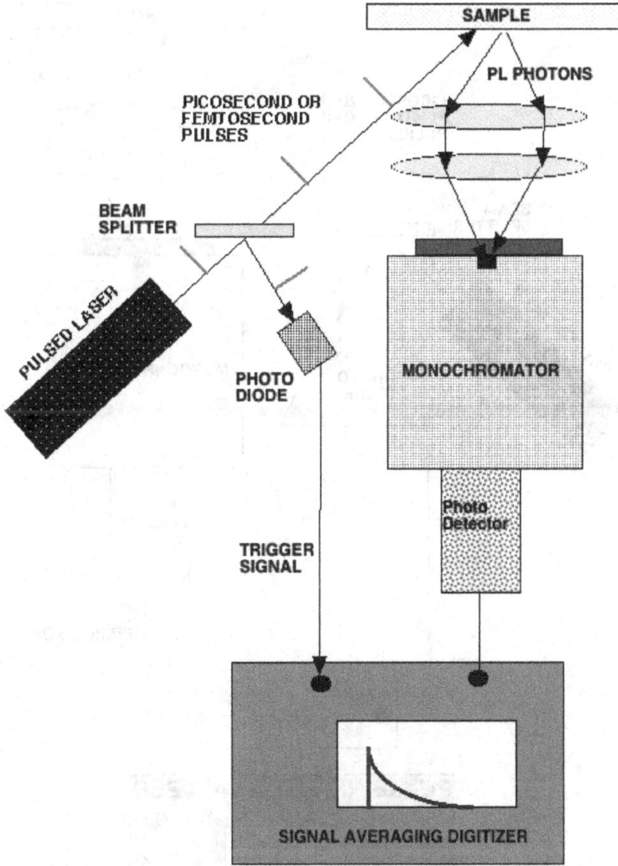

Fig. 7.1. Time-resolved photoluminescence measurement apparatus using analog detection.

Comparison of simultaneous measurements provide great detail about the recombination process.

7.1. Time-correlated Single Photon Counting (TCSPC)

TCSPC is the most widely used and sensitive to measure the lifetime of carriers in direct bandgap materials. A typical apparatus configuration is shown below:

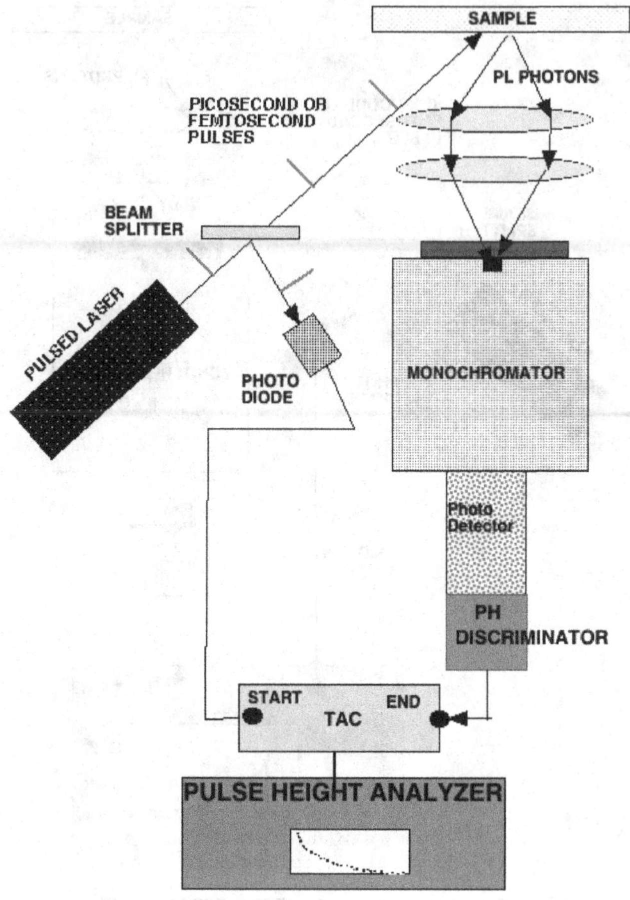

Fig. 7.2. Time-correlated single photon counting system.

The time-correlated single photon counting method (TCSPC) has been the standard for several decades[3] and is currently the standard method to measure direct bandgap materials. TCSPC measurement systems are currently commercially available. The photon counting technique was developed for semiconductors by Bachrach,[4] and has become a standard over the years. The first formal description was provided in the book by Demas[5] and was originally applied to research in the chemistry community. NREL adopted the technique for photovoltaic materials about 1985. The original NREL facility was adopted by NREL[6] in 1988. A later publication described the measurement theory in more detail.

In order to use time-correlated single photon counting, the photodetector must have sufficient gain to detect single photons produced by the recombination free electron-hole pairs. There are numerous commercial photodetectors with sufficient gain for single photon counting. These include photomultiplier tubes (PMT), microchannel plates (MCP), and diode detectors for the infrared spectral region. The latter were first described by Louis[7] and used an avalanche photodiode (SPAD) for photon counting in semiconductors. Current detector technology for the near-infrared region has allowed extension of TRPL photon counting measurements to small bandgap semiconductors ($E_g < 1.1\,\mathrm{eV}$).

The research community has numerous systems that use a variety of modern components. The original system used a S-1 photomultiplier tube had an impulse response of about 300 picoseconds (ps). The relatively long impulse response is caused by the transit time-dispersion of electrons as they cascade down the dynode chain. Transit time dispersion is greatly reduced with new detector technology. One of the later innovations was a microchannel plate detector. The first microchannel plate detectors that had a time resolution of 30 ps full-width half-maximum (FWHM). The time-correlated photon counting electronics are similar for the remainder of the detection system.

A schematic of typical setup is shown in Fig. 7.2. The photons emitted by the sample are focused on the input slit of a scanning monochromator. The monochromator is tuned to a narrow band of wavelengths in the intrinsic or band-to-band emission spectra of the semiconductor. Using a beam splitter and a photodiode, a small fraction of the incident laser pulse is deflected to a fast photodiode for initiating a timing pulse. The current technology uses laser pulses from picosecond to femtosecond full width half maximum (FWHM). The electrical output of the photodiode triggers a sweep of the time-to-amplitude converter (TAC). The TAC is a small electronic component that executes a time sweep over the desired time span. The sweep time is set on the TAC panel and tailored to the expected lifetime of the sample. The pulse-height discriminator is necessary to block electrical pulses that are produced by thermal and other non–single-photon events. The first collected photon initiates an electrical pulse in the photodetector that is amplified and passed through an amplitude discriminator. The electrical pulse produces a stop message at the TAC. The TAC generates a voltage ramp that is terminated by the arrival of the first collected photon. Therefore, the ramp amplitude proportional to the time delay between the trigger pulse and photon arrival. A TAC signal is directed to a multi-channel pulse height analyzer (MCA) and produces one count in the appropriate time-delay channel. Thus, the signal from each photon is stored in a channel appropriate to the time delay. One count is recorded for every collected photon pulse. The output of the TAC is a histogram of the PL decay that is composed of photon counts versus time.

A TCSPC system typically collects a PL photon every 100 to 1000 laser excitation pulses. Therefore, a fairly high laser repetition rate is desirable for rapid data acquisition using TCSPC. At the other extreme, the repetition period (1/f) must be larger than the sample decay time. Otherwise, overlap of decay events produces errors in the TAC operation. For example, a pulse repetition rate of 100 KHz is usually adequate for doped GaAs with a decay time much less than 10 μs. A pulse repetition rate of 1 MHz is suitable for thin film polycrystalline materials such as CdTe with typical ns lifetimes.

Single photon counting is a very sensitive technique, compared to the analog method, and requires very low light levels. PL decay measurements can be made at very low injection levels because of this sensitivity. Detectable measurements of PL decay can be made at excitation levels as low as 10^{-4} μ-Joules/cm^2. Photon counting also has a wide dynamic range so that excitation levels can range from very low to very high injection.

Materials with low doping density and long lifetime require many minutes of data accumulation time to obtain a significant number of photon counts as was the case for the sample data shown here. The more efficient method of measuring these devices is to use the photoconductive decay such as the RCPCD method. In that case, quality data is produced by several seconds of measurement.

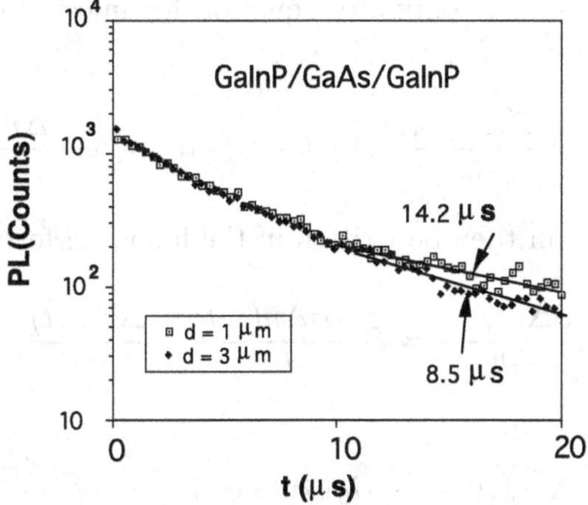

Fig. 7.3. TCSPC of GaInP/GaAs DH with active layer thicknesses shown. The GaAs was not intentionally doped but the background doping is estimated at 2–5 $\times$ 10^{14} cm^{-3}. [Reprinted from Olson, J. M., Ahrenkiel, R. K., Dunlavy, D. J., Keyes, B. M., and Kibbler, A. E. (1989). *Appl. Phys. Lett.* **55**, 1208, with permission of AIP Publishing.]

To get an accurate measurement of the SRV of the DH interface, one grows a nominally undoped active layer to minimize the bulk radiative recombination rate and forces the interface to be the controlling recombination mechanism. The NREL group used TCSPC measurements to measure the first high-quality, undoped GaInP/GaAs DHs grown by Olson and coworkers in their laboratories.[8] The GaAs layer was not intentionally doped and has a background doping density of about $5 \times 10^{13}\,\text{cm}^{-3}$. Therefore, the low SRV controls the lifetime and is estimated at less than $5\,\text{cm/s}$.

7.2. Radiative Recombinaton at High Injection

This optical pulse generates an instantaneous concentration of electron-hole pairs in a Beers law distribution that varies with the wavelength of the monochromatic light. We will analyze the case of thin film, p-type material that has a thickness of submicron to several microns. The nearly instantaneous generation of electron-hole pairs at $t = 0$, is described by an impulse function, $\delta(t)$ in the analysis.

The one-dimensional continuity equation for injected electrons can be written as:

$$\frac{\partial \Delta n(x,t)}{\partial t} = D_n \frac{\partial^2 \Delta n(x,t)}{\partial x^2} + \alpha I_0 \exp(-\alpha x)\delta(t) - \frac{\Delta n(x,t)}{\tau}. \quad (7.2)$$

This equation can then be written in the following form for $t > 0$:

$$\frac{\partial \Delta n(x,t)}{\partial t} = D_n \frac{\partial^2 \Delta n(x,t)}{\partial x^2} - \frac{\Delta n(x,t)}{\tau}$$

with

$$\Delta n(x,0) = +\alpha I_0 \exp(-\alpha x) \quad (7.3)$$

Here τ is the recombination lifetime, which was assumed to be constant in the prior analysis of planar structures. As before, D_n is the minority electron diffusivity and α is the absorption coefficient of the monochromatic light. We will modify Eq. (5.15), with the substitution of the injection-dependent radiative recombination rate for the constant bulk lifetime rate of $1/\tau$. Thus:

$$\Delta n(x,t) = A_n \exp(-B(\Delta n N_{\mathrm{A}} + \Delta n^2)t \sum_{n=1}^{\infty} \exp(-\beta_n t) \times \left[\cos(\alpha_n x) + \frac{S}{\alpha_n D}\sin(\alpha_n x)\right]. \quad (7.4)$$

From the previous analysis, we know that the Fourier modes decay rapidly with increasing n-terms in the expansion. From Fig. 5.1, one sees that:

$$\alpha_1 < 1/2\frac{\pi}{d};$$

$$\alpha_n \sim \alpha_1 + n\frac{\pi}{d}. \quad (7.5)$$

For a thin film of micron thickness, d, the argument α_n increases rapidly with term number. The time decay argument, β_n, increases as the square of α_n. Thus:

$$\beta_1 = D\alpha_1^2;$$
$$\beta_n \sim D\left(\alpha_1 + (n-1)\frac{\pi}{d}\right)^2. \quad (7.6)$$

For thin films, the higher modes decay rapidly and the the first mode dominates the injected electron density function.

Here, we assume a symmetrical structure with $S_1 = S_2$ and A_1 is the amplitude of the first mode. Thus, for thin films of direct bandgap material, one can approximate the decay time as:

$$\Delta n(x,t) = \exp(-B(\Delta n N_A + \Delta n^2)t^* \exp(-\beta_1 t)^* A_1$$
$$\times \left[\cos(\alpha_1 x) + \frac{S}{\alpha_1 D}\sin(\alpha_1 x)\right]. \quad (7.7)$$

This expression is accurate for times larger than:

$$t > 1/\beta_1.$$

For longer times, the excess electron "flattens" to that of the first mode as the higher modes decay. When $\Delta n > N_A$, we see that the instantaneous decay time is initially much shorter than the low-injection radiative lifetime. At times longer than the radiative lifetime, bulk decay time becomes the radiative lifetime, τ_R. When the SRV parameter $Z = Sd/D$ is small ($Z \ll 1$), the solution of the eigenvalue equation is approximately the following:

$$\tan(\theta_1) = \frac{2Z\theta_1}{\theta_1^2 - Z^2} \sim \frac{2Z}{\theta_1}. \quad (7.8)$$

where,

$\theta_1 = \alpha_1 d$. The result is that β_1 is calculated as:

$$\alpha_1^2 = \frac{2S}{Dd};$$
$$\beta_1 = D\alpha_1^2 = \frac{2S}{d}. \quad (7.9)$$

The time dependence of the electron density is (for small S values):

$$\Delta n(x,t) = \exp(-B(\Delta n N_A + \Delta n^2)t^* \exp\left(-\frac{2S}{d}t\right)^* A_1 \times \left[\cos(\alpha_1 x) + \frac{S}{\alpha_1 D}\sin(\alpha_1 x)\right] \quad (7.10)$$

and

$$\alpha_1 = \sqrt{\frac{2S}{Dd}}.$$

In the limit of $S = 0$, the decay is purely radiative and $\Delta n(x,t)$ becomes:

$$\Delta n(x,t) = A_1 \exp(-B(\Delta n N_A + \Delta n^2)t. \quad (7.11)$$

The total initial electron concentration is calculated by integrating the optical generation function across the volume of the material.

$$\Delta n_0 = \int_0^d \alpha I_0 \exp(-\alpha x)dx = I_0(1 - \exp(-\alpha d)). \quad (7.12)$$

The time dependence of the carrier decay that is dominated by radiative recombination is:

$$\frac{d\Delta n}{dt} = -B(\Delta n p_0 + \Delta n^2). \quad (7.13)$$

A closed form solution of the above is:

$$\Delta n(t) = \Delta n_0 \frac{\exp(-t/\tau_R)}{1 + \frac{\Delta n_0}{p_0}(1 - \exp(-t/\tau_R))}. \quad (7.14)$$

The rate of decay is nonexponential until $t \gg \tau_R$. At that time, the excess electrons decay at the low-injection radiative lifetime rate:

$$\Delta n(t) = \Delta n_0 \exp(-t/\tau_R) \text{ where } \tau_R \text{ is radiative lifetime.}$$

At any injection level, the PL decay can be written as:

$$I_{PL}(t) = B(p_0 \Delta n(t) + \Delta n(t)^2). \quad (7.15)$$

7.3. TCSPC Data Measured on Doped GaAs

TCSPC data is shown in Fig. 7.4 for a *p*-AlGaAs/GaAs DH grown by MOCVD and doped to a level of 1×10^{17} cm^{-3}. The DH thickness is 10 μm and the photon recycling effects has increased the low injection lifetime by about a factor of ten. The measured low injection lifetime (curve B) is 532 ns and the radiative lifetime is about 50 ns. The measured injection level, Δn, is about 2×10^{15} cm^{-3} as measured by a calibrated energy meter using the measured pulse repetition rate. The injected energy density is calculated assuming that the initial surface electron areal density diffuses through the volume before significant recombination occurs. In curve A, the injection level increases to about 1×10^{18} cm^{-3}, producing a much faster initial decay rate. The measured initial lifetime is 32 ns, which is about a factor of ten larger than the calculated high injection radiative lifetime of 2.5 ns. The photon recycling effect plays a role in producing this larger high-injection lifetime. These data illustrate the effects of high injection on the decay curve. To measure the radiative lifetime, one must ignore the initial fast decay or reduce the intensity until low injection behavior is observed. This behavior of the decay

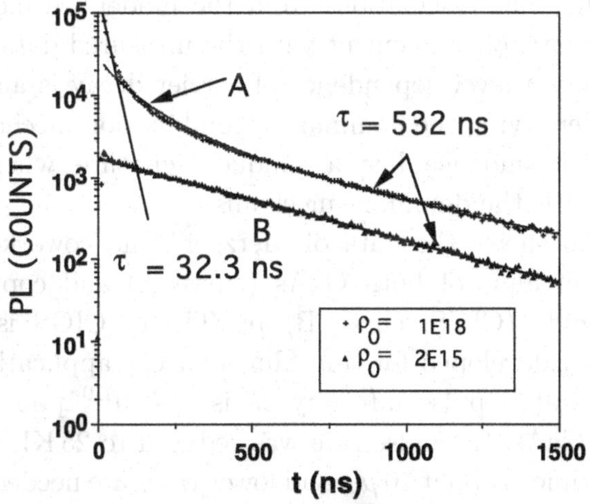

Fig. 7.4. The TRPL (or TCSPC) decay curves for a *p*-GaAs DH doped to 1×10^{17} cm^{-3} measured at high injection (A) and low injection (B).

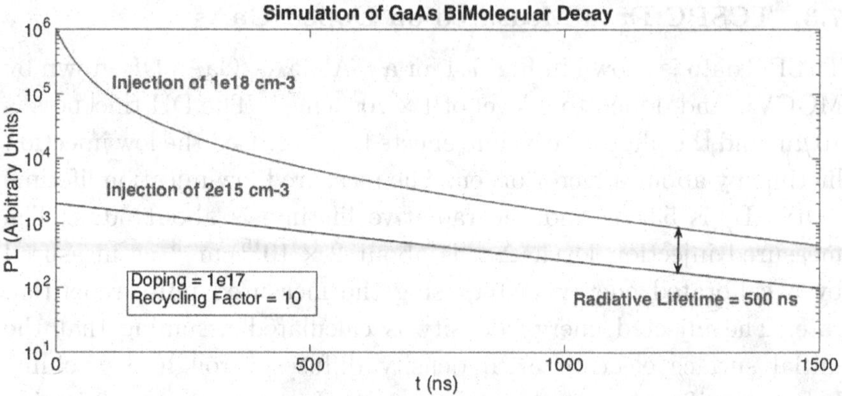

Fig. 7.5. The bimolecular decay curves using the parameters for the GaAs DH device of Fig. 7.4 using Eq. (7.15). The comparison with the data is quite good.

time with pulse intensity is a good indicator of dominant radiative recombination.

These data are successfully modeled using equation (7.14) and shown below. Here the models use a film thickness of $10\,\mu$m, and a calculated photon recycling factor of ten. The film doping used in the model was $1 \times 10^{17}\,\text{cm}^{-3}$. These data were simulated by a model that used the bimolecular expression and the parameters of the GaAs DH. The calculations from the model are plotted below and show remarkable agreement with the measured data.

The injection level dependence of carrier decay is an important method of identifying the dominant recombination mechanism. This bimolecular dependence has a unique signature when radiative recombination is the dominant mechanism.

Figure 7.6 shows the data of Metzger[9] and coworkers as they measured thin films of both GaAs (curve A) and copper indium gallium selenide (CIGS, curve B) by TRPL. CIGS is a current material being developed for thin film solar cell applications.

The calibrated pulse intensity I_0 is 4×10^{10} photons/cm^2 at 680 nm. For GaAs, the pulse rate was reduced to 25 KHz as the low injection lifetime is about $10\,\mu$s, and lower rates are needed to prevent decay overlap. Metzger *et al.*[9] discuss the surface properties of CIGS. As grown, the surface appears to be self-passivated, but surface states develop after exposure to air in times from minutes to hours.

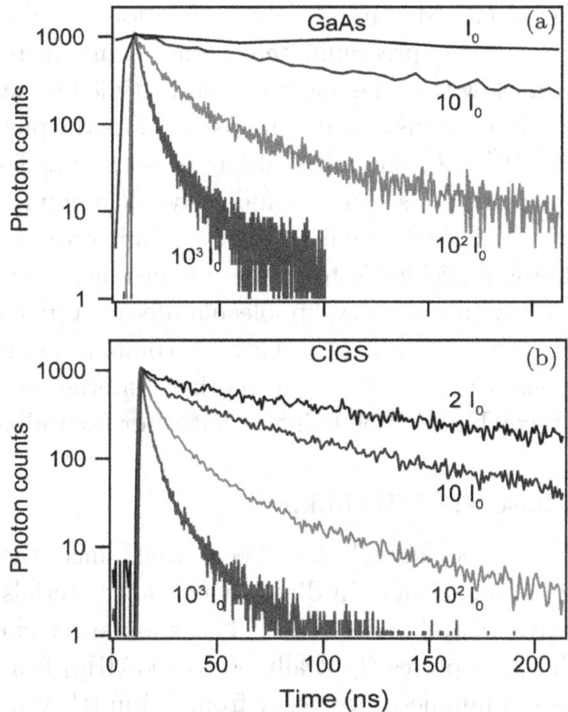

Fig. 7.6. The injection level dependence of GaAs (a) and CIGS (b) thin films using TRPL as measured by Metzger and coworkers. The TCSPC data was generated by a 670 nm pulsed laser pulse. The flux level I_0 is 4×10^{10} photons/cm^2 at 680 nm. [Reprinted from *Thin Solid Films* **517**, Metzger, W. K., Repins, I. L., Romero, M., Dippo, P., Contreras, M., Noufi, R., and Levi, D., Recombination kinetics and stability in polycrystalline Cu(In,Ga)Se2 solar cells, 2360–2364 (2009) with permission from Elsevier.]

A thin layer of CdS preserves the passivation for many months. The there is a background low injection lifetime of 100 to 150 ns for the as-grown films, that appears to be low-level defect recombination. The background self-doping in the range 5×10^{15} to 2×10^{16} cm^{-3} and is *p*-type. The GaAs is a standard double heterostructure with a thickness, *d*, of about 0.75 μm. The background self-doping in such samples is typically 5×10^{14} cm^{-3} or less. The latter value gives a low injection radiative lifetime of about 10 μs. The decay behavior varies from exponential at the lowest injection levels to bimolecular at the higher injection levels, which is increased in steps to $10^3 \times I_0$.

The decay behaviors identify the recombination mechanism in the two thin films as being predominately radiative in nature.

These data show that the low-injection lifetime for GaAs is 10 μs and that of CIGS is 100 ns at pulse intensity I_0. As the pulse intensity is increased to $10^3 \times I_0$, the initial decay of both films is nearly the same at about 2 ns. This result is indicative of dominant radiative recombination. Also the B-coefficient is very similar for the two materials. The photon recycling factor is an unknown here but is minimal for very thin films. In summary, bimolecular decay at high injection is a reliable indicator of dominant radiative recombination mechanism. Bimolecular behavior can be sought in other materials as a means to identify the recombination as being radiative or nonradiative.

7.4. The Phase Shift Technique

The phase shift technique[10] has been used since the 1970s to measure the carrier lifetime in direct-band gap materials that have strong radiative photoluminescence. The standard technique used high repetition rate pulses (typically 50 to 100 MHz) from a laser to produce pulses of luminescence decay from a thin III–V film. Using a fast photo-detector to capture the PL pulses, the phase shift between the excitation and PL pulses was proportional to the carrier lifetime. This system was used at NREL in the 1980s to measure the carrier lifetime in thin films. The data analysis[11] was discussed in detail in an article by the first author (RKA) in the "Current Topic in Photovoltaics" series.

Using the continuity equation for a material in a confinement structure with a sinusoidal optical input, the equation is written:

$$\frac{\partial \Delta n(t)}{\partial t} = G_0 \exp(i\omega t) - \frac{\Delta n(t)}{\tau}; \tag{7.16}$$

Here, G_0 is the amplitude of the sinusoidal light source.

Using a trial solution: where ϕ is an unknown phase shift.

$$\begin{aligned} \Delta n(t) &= \Delta n_0 \exp(i\omega t - \varphi); \\ \frac{\partial n(t)}{\partial t} &= i\omega \Delta n(t) = G_0 \exp(i\omega t) - \frac{\Delta n(t)}{\tau}; \end{aligned} \tag{7.17}$$

Inserting the trial solution and cancelling the common time dependence term:

$$\Delta n_0 = \frac{G_0 \exp(i\varphi)}{\frac{1}{\tau} + i\omega}; \quad \text{or simplifying:} \tag{7.18}$$

$$\Delta n_0 = \frac{\tau G_0 \exp(i\omega\phi)}{1 + i\omega\tau}; \quad \text{Then;}$$

$$\Delta n(t) = G_0 \tau \frac{1 - i\omega\tau}{1 + \omega^2\tau^2} \exp(i\omega t); \tag{7.19}$$

This can be written as a phase shift of the carrier density.

$$\Delta n(t) = \frac{G_0 \tau}{\sqrt{1 + \omega^2\tau^2}} \exp(-i\theta) \exp(i\omega t); \tag{7.20}$$

Here:

$$\tan(\theta) = \omega\tau;$$

where τ is the carrier lifetime and ω is the frequency of the light source. The measurement of the phase shift between light source and the detected PL signal provides a lifetime value. For a PL signal, the modulated photoluminescence is expressed by:

$$I_{\text{PL}}(t) = \frac{BNG_0 \tau}{\sqrt{1 + \omega^2\tau^2}} \exp(-i\theta) \exp(i\omega t); \tag{7.21}$$

Here, B is the radiative coefficient of silicon and N is the doping level. This excludes the Δn^2 term of high injection. Thus, one gets an absolute lifetime value that is independent of the other parameters of the measurement. This can be performed as a function of injection level for a second parameter and give additional information.

This technique is applicable when the decay time, τ, is much shorter than the period of the excitation source, T, where: $T = \frac{2\pi}{\omega}$; For the mode locked laser excitation source used in the original configuration and repetition rates of 100 MHz, one is limited to lifetimes of less than 100 ns. This configuration is obviously inappropriate for silicon.

However, Giesecke[12] and coworkers used a 880 nm diode laser for low frequency excitation of silicon wafers grown by the float zone technique. Their modulation frequency varied from 1.0 to 300 Hz and the maximum incident intensity varied from 10^{16} to $10^{19}\,cm^{-2}\,s^{-1}$. They were able to measure lifetimes as short as $20\,\mu s$ using phase shift information.

7.5. Upconversion

The standard photon counting technique fails for several special situations. For small bandgap materials, the technology lacks a detector that is sufficiently sensitive that allows single photon counting. For very short lifetime materials, the standard detectors do not have a response time that allows for the observation of very fast events; especially in the sub-picosecond time domain. For such measurements, a novel technique called upconversion has been used for a number of years.[13] In this technique, the PL decay photons are mixed with a time delayed portion of the excitation pulse in a nonlinear crystal. The crystal generates optical pulses that are the sum and difference of the incident photons. The upconverted photons, are detected by a standard fast detector.

$$h\nu_{sum} = h\nu_{PL} + h\nu_{laser}. \tag{7.22}$$

The detector process does not require time resolution of the PL photon, but merely the presence or absence of the later. The time delay is provided by the optical delay of the laser pulse by using the time-of-flight time as the metric of resolved the arrival time of the PL pulse relative to the laser excitation pulse. The apparatus constructed by Ellingson[14] and used with coworkers as shown in Fig. 7.7.

Samples in the higher doping or small bandgap ranges were measured by the up conversion photoluminescence photon-counting technique. The samples were pumped by 90 femtosecond (fs) pulses at 82 MHz from a Spectra-Physics femtosecond modelocked titanium-doped sapphire (Ti:S) laser running at 762 nm. The collected luminescence was mixed with the Ti:S laser pulse for sum-frequency generation (SFG) in a 1-mm lithium iodate ($LiIO_3$) crystal, and the SFG signal was dispersed by a SPEX 270 m spectrograph and detected by a Hamamatsu R464 photomultiplier tube. The system

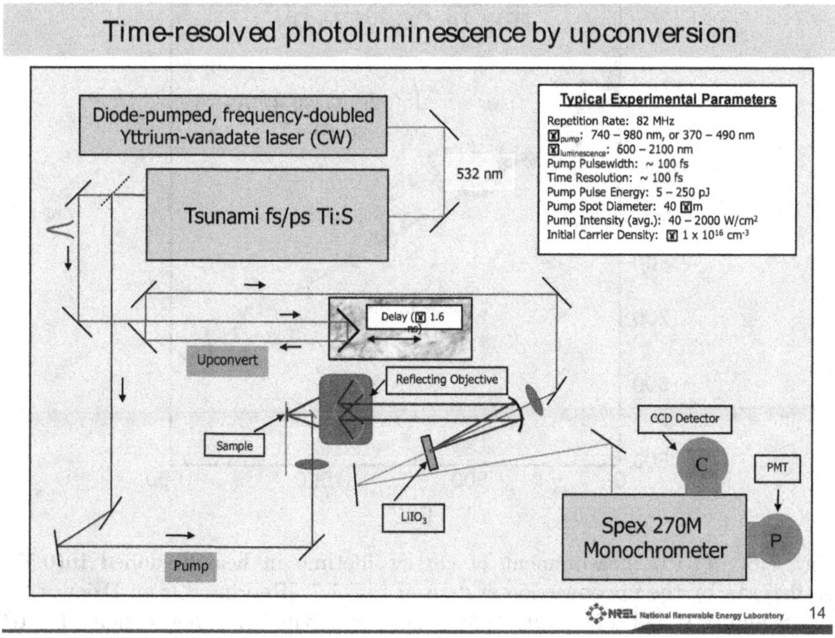

Fig. 7.7. The upconversion system developed and constructed by R. Ellingson for the measurements described here.

time resolution is approximately 110 fs, and the detection wavelength of 526 nm corresponds to an In(0.53)Ga(.47)As luminescence peaking at 1.68 um.

The data in Fig. 7.8 were obtained on a double heterostructure of lattice-matched composition InP/In(0.53)As(0.47)As/InP grown by MOCVD. The InGaAs was doped n-type to a level 6×10^{17} cm^{-3}. The thickness of the active layer is 3.0 μm. The data showed a single exponential decay with a lifetime 2.43 ns.

The upconversion allows measurement of lifetimes in the subnanosecond time domain. Data in the following chapters will use data from this apparatus to measure the PL lifetime of small bandgap InGaAs alloys and heavily doped GaAs.

Data were obtained from a concentration series of In(0.53)As (0.47)As as a means to measure the radiative and Auger coefficients. The data for 6×10^{17} cm^{-3} are shown in Fig. 7.8. A number of

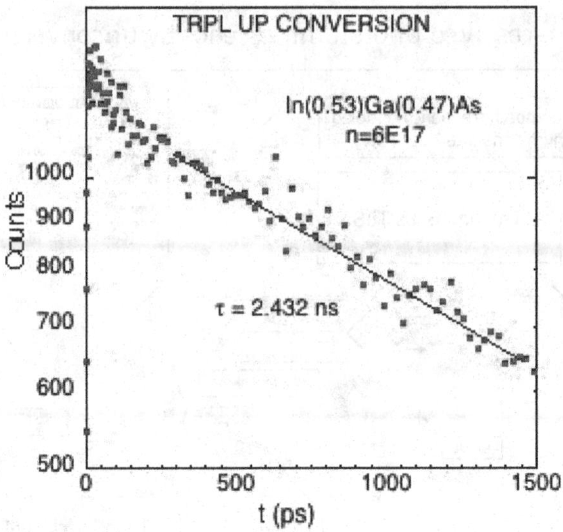

Fig. 7.8. TRPL measurement of carrier lifetime in heavily doped In(0.53) Ga(0.47)As by the upconversion system of Fig. 7.7. [Reprinted from Dziewor, J. and Schmid, W. (1977). *Appl. Phys. Lett.* **31**, 346, with permission of AIP Publishing.]

other measurements were made on small bandgap and heavily doped semiconductors.

7.6. TRPL on Single Crystal Silicon

In 1977, Dziewior and Schmid[15] used TCSPC to measure the lifetime of doped n- and p-type silicon in the Auger range with $N_{\mathrm{A,D}}$ greater then $1 \times 10^{17}\,\mathrm{cm}^{-3}$. As the quantum efficiency is very low in this indirect bandgap material, they used an S-1 photomultiplier cooled to liquid nitrogen temperature for detection. They used a single photon technique with a TAC similar to that described earlier in this section. The light source was a mode-locked cavity-dumped Ar+-ion laser operating at 10 kHz to 1 MHz corresponding to the time needed for decay and recovery. Their data is shown in Fig. 7.9. The detection system was tuned to 1.1 μm and the laser output was tuned to 514.5 nm.

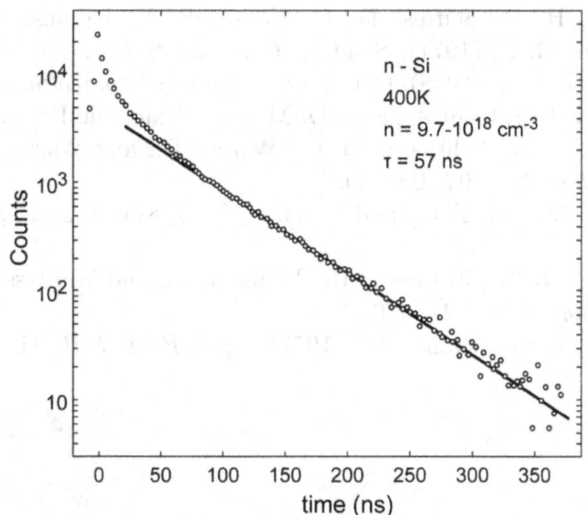

Fig. 7.9. The photoluminescence decay of n-type silicon obtained by Dziewior and Schmid at 400 K by TCSPC.

The sample temperature here was 400 K but data were measured at 77 K, 300 K, and 400 K. Clearly, high quality data were obtained by this technique. The resulting Auger coefficients for n- and p-type silicon will be discussed in Chapter 8.

References

1. Ahrenkiel, R. K. (1993). *Semiconductors and Semimetals*, Academic, New York, Vol. 39, pp. 39–150.
2. Maiberg, M. and Scheer, R. (2014). *J. Appl. Phys.* **116**, 123710.
3. Bollinger, L. M. and Thomas, G. E. (1961). *Rev. Sci. lnstrum.* **32**, 1044.
4. Bachrach, R. Z. (1972). *Rev. Scientific Instruments* **43**, 734.
5. Demas, J. N. (1983). In *Excited State Lifetime Measurements*, Academic Press, New York.
6. Ahrenkiel, R. K. (1992). *Solid State Electronics* **35**, 239.
7. Louis, T. A., Ripamonti, G. and Lacaita, A. (1990). *Rev. Sci. Instrum.* **611**, 11.
8. Olson, J. M., Ahrenkiel, R. K., Dunlavy, D. J., Keyes, B. M., and Kibbler, A. E. (1989). *Appl. Phys. Lett.* **55**, 1208.
9. Metzger, W. K., Repins, I. L., Romero, M., Dippo, P., Contreras, M., Noufi, R., and Levi, D. (2009). *Thin Solid Films* **517,** 2360–2364.

10. Zwicker, H. R. Scifres, D. R. Holonyak, N. Dupuis, R. D., and Burnham, R. D. (1971). *Solid St. Commun*, **9**, 587.
11. Ahrenkiel, R. K. (1988). In *Current Topics in Photovoltaics, Volume 3*, edited by T. J. Coutts and J. D. Meakin, Academic Press, pp. 44–50.
12. Giesecke, J. A., Schubert, M. C., Walter, D., and Warta, W. (2010). *Appl. Phys. Lett.* **97**, 092109.
13. Shah, J., Damen, T. C., and Deveaud, B. (1987). *Appl. Phys. Lett.* **50**, 1307.
14. Ahrenkiel, R. K., Ellingson, R., Johnston, S., and Wanlass, M. (1980). *Appl. Phys. Lett.* **72**, 3470.
15. Dziewor, J. and Schmid, W. (1977). *Appl. Phys. Lett.* **31**, 346.

CHAPTER 8

Auger Recombination

8.1. Recombination Kinetics

Auger recombination is a nonradiative recombination process in which the energy of the electron-hole pair is transformed into the kinetic energy of a free particle. The process is the inverse of impact ionization. In this case, an energetic minority carrier recombines with a majority carrier and transfers the excess energy to another majority carrier. A large variety of Auger processes that involve not only the free electron and hole, but also involve phonons and localized states (traps). A compilation of the various Auger processes has been described in an early review by Landsberg and Robbins.[1] The minority-carrier lifetime in undoped crystalline silicon is dominated by Auger recombination at injection levels exceeding[2] about $1 \times 10^{17}\ \mathrm{cm}^{-3}$. The recombination mechanisms of undoped silicon, over a range of injection levels, are described by a combination of SRH, band-to-band Auger, and trap-Auger effects.[3]

Auger recombination is often a limiting factor at the high injection levels in LEDs and solid state lasers. The recombination rate varies with n^2p in n-type or np^2 in p-type nondegenerate semiconductors at relatively high carrier concentrations. The complicating effects of degeneracy on Auger recombination rates are discussed by Landsberg.[4] The three-carrier Auger recombination lifetime, τ_a, for

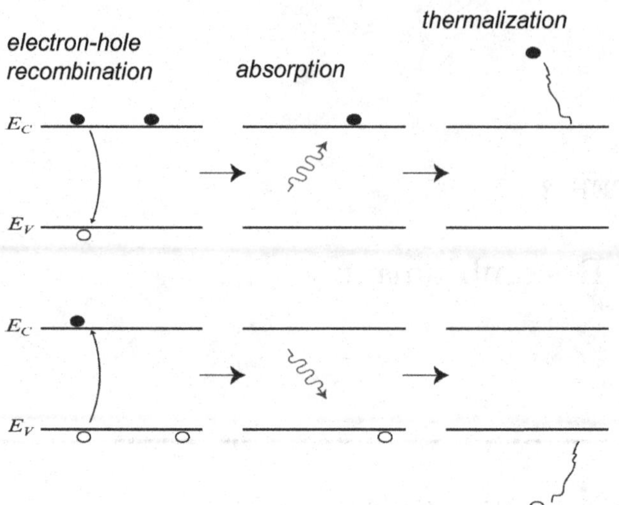

Fig. 8.1. Schematic representation of Auger recombination processes for electrons and holes.

n and p-type semiconductors varies as[5]:

$$\begin{aligned} \frac{1}{\tau_a} &= Anp + Cn^2 \quad (n\text{-type}) \\ \frac{1}{\tau_a} &= Anp + Cp^2 \quad (p\text{-type}) \end{aligned} \tag{8.1}$$

Here the coefficient C is a cross section that is proportional to the transfer efficiency of energy to the majority carrier. The A coefficient represents the transfer of energy to the minority carrier.

Auger processes involve the intraband exchanges of energy. For example, a photon spontaneously emitted by an electron during recombination with a hole can be absorbed by another electron, which is elevated above the CB edge

The absorbed energy is transformed to heat during thermalization to the band edge. The concentration dependence of the recombination rates is written:

$$\begin{aligned} U_{p,\mathrm{Aug}} &= C_p \cdot (n^2 \cdot p - n_0^2 \cdot p_0), \\ U_{n,\mathrm{Aug}} &= C_n \cdot (n \cdot p^2 - n_0 \cdot p_0^2). \end{aligned} \tag{8.2}$$

Because of the quadratic dependence on one carrier concentration, the disparity in Auger rates between minority and majority carriers is even more pronounced than for other mechanisms. In the case of a p-type material with $p = N_a$, the Auger recombination rate for electrons is:

$$\begin{aligned} U_A &= C_n \cdot [(n_0 + \Delta n) \cdot N_a^2 - n_0 \cdot p_0^2] \\ &\approx C_n \cdot N_a^2 \cdot \Delta n = \frac{\Delta n}{\tau_{n,\text{Aug}}} \end{aligned} \tag{8.3}$$

where we can identifiy the Auger lifetime:

$$\frac{1}{\tau_{n,\text{Aug}}} = C_n \cdot N_a^2. \tag{8.4}$$

Specific data on Auger recombination rates in various materials will be discussed in later in this section.

In direct bandgap materials, there may a doping region in which both radiative and Auger recombination are significant. One can define a crossover region (N_c) in which the two recombination rates are equivalent. Or:

$$\begin{aligned} BN_c &= CN_c^2 \\ N_c &= B/C. \end{aligned} \tag{8.5}$$

where N_c is the crossover carrier concentration.

Samples in the higher doping ranges were measured by photoluminescence photon-counting technique. The sum-frequency upconversion technique is required for very short lifetimes that are found in the Auger-controlled concentration of doped and/or injected carriers. The samples were pumped by 90 femtosecond (fs) pulses at 82 MHz from a Spectra-Physics femtosecond mode-locked titanium-doped sapphire (Ti:S) laser running at 762 nm (1.627 eV). The collected luminescence was mixed with the Ti:S laser pulse for sum-frequency generation (SFG) in a 1-mm lithium iodate ($LiIO_3$) crystal, and the SFG signal was dispersed by a SPEX 270 m spectrograph and detected by a Hamamatsu R464 photomultiplier tube. The system time resolution is approximately 110 fs. The detection wavelength of

λ is selected to match the bandgap of the selected material, where the detection wavelength is;

$$h\nu_{\mathrm{PM}} = 1.627 + E_g. \tag{8.6}$$

$$\lambda_{\mathrm{PM}}(\mathrm{nm}) = \frac{1240}{h\nu_{\mathrm{PM}}}.$$

8.2. Auger Data

8.2.1. *Silicon*

Early experimental work[6] found that the carrier lifetime in both n- and p-type silicon was dominated by Auger recombination at concentrations levels above $1 \times 10^{18}\,\mathrm{cm}^{-3}$. The PL signal was excited by a mode-locked Ar+ laser with a 400 ps pulse width. Detection was accomplished by single-photon counting. The lifetime versus concentration at 77 K, 300 K, and 400 K temperatures. The data for p-type silicon is shown in Fig. 8.2(a) and that of n-type is shown in Fig. 8.2(b).

The experimental work by Dziewior[6] and Schmid found that the carrier lifetime in both n- and p-type silicon was dominated by Auger recombination at concentrations levels above $1 \times 10^{18}\,\mathrm{cm}^{-3}$. The Auger coefficient was measured by these workers by TCSPC, as described in the previous chapter, at 77 K, 300 K, and 400 K. These measurements were made on doped float zone wafers with P and B. The data at 77 K, 300 K, 400 K as a function of doping level are shown below:

The Auger coefficients that are calculated from the PL lifetime fits below.

$$C_n = 2.8 \times 10^{-31}\,\mathrm{cm}^6\,\mathrm{s}^{-1}.$$
$$Cp = 9.9 \times 10^{-32}\,\mathrm{cm}^6\,\mathrm{s}^{-1}. \tag{8.7}$$

Therefore, the electron Auger coefficient is 2.8 times larger than the hole coefficient. The crossover concentrations for silicon are:

$$N_{cn} = B/C_n = 7 \times 10^{15}\,\mathrm{cm}^{-3},$$
$$N_{cp} = B/C_p = 2 \times 10^{16}\,\mathrm{cm}^{-3}. \tag{8.8}$$

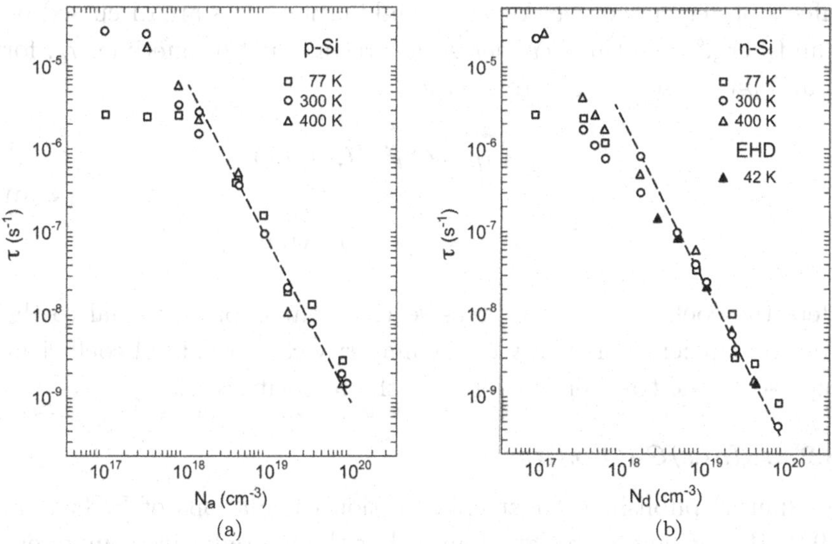

Fig. 8.2. Carrier lifetime[6] at high doping levels measured on (a) p-type silicon by TCSPC at three temperatures and (b) n-type silicon by TCSPC at three temperatures. [Reprinted from Dziewor, J. and Schmid, W. (1977). *Appl. Phys. Lett.* **31**, 346, with permission of AIP Publishing.]

The crossover region for silicon is in a doping region that is usually dominated by SRH or defect recombination. Therefore, one is more likely to see a transition from SRH to Auger recombination without a significant contribution from radiative recombination.

Pang and Rohatgi[7] introduced a new method to analyze photoconductive decay data to separated SRH. radiative, and Auger processes. Using their fitting program, they obtained the following results for an oxidized CZ wafer at over a range of injection levels.

$$B = 3.0 \times 10^{-15}\,\text{cm}^3/\text{s}; \qquad C = 1.1 \times 10^{-30}\,\text{cm}^6/\text{s}. \tag{8.9}$$

Auger recombination is often a limiting factor at the high injection levels in LEDs and solid state lasers. The recombination rate varies with n^2p in n-type or np^2 in p-type nondegenerate semiconductors at relatively high carrier concentrations. The complicating effects of degeneracy on Auger recombination rates are discussed by Landsberg.[4] The three-carrier Auger recombination lifetime, τ_a, for n and p-type semiconductors varies as[5]

$$\begin{aligned} \frac{1}{\tau_a} &= Anp + Cn^2 \quad (n\text{-type}) \\ \frac{1}{\tau_a} &= Anp + Cp^2 \quad (p\text{-type}) \end{aligned} \tag{8.10}$$

Here the coefficient C is a cross section that is proportional to the transfer efficiency of energy to the majority carrier. The A coefficient represents the transfer of energy to the minority carrier.

8.2.2. *In(x)Ga(1-x)As*

Sugimura[8] published Auger cross-section calculations of InGaAs in 1981. Henry[9] and coworkers found that the nonradiative component of the minority-carrier lifetime in InGaAs was dominated by Auger recombination. The Auger cross section is significantly larger in p-type than in n-type InGaAs layers. Larger Auger effects are seen in the quaternary semiconductor InGaAsP by Henry[10] and coworkers.

The time-resolved PL measurements by the first author[11] and coworkers of In(0.53)Ga(.47)As were fit by an Auger coefficient that were similar for n- and p-type materials.

$$C_n = C_p = 8.1 \times 10^{-29}\,\text{cm}^6\,\text{s}^{-1}. \tag{8.11}$$

As seen in Fig. 7.8, there is considerable scatter in the data. The crossover concentration from radiative to Auger is:

$$N_c = B/C = 1.76 \times 10^{19}\,\text{cm}^{-3}. \tag{8.12}$$

Thus, radiative recombination dominates over a large portion of the common doping ranges of this composition.

Follow on work by Metzger[12] and coworkers measured carrier lifetime in the lattice-matched ($x = 0.74$) and lower bandgap n-type In(x)Ga($1-x$)As, again using the up-conversion technique. Figure 8.3 shows lifetime data on compositions resulting in $Eg = 0.74$, 0.60, and 0.50 eV. The change of slope at the higher doping levels is evident

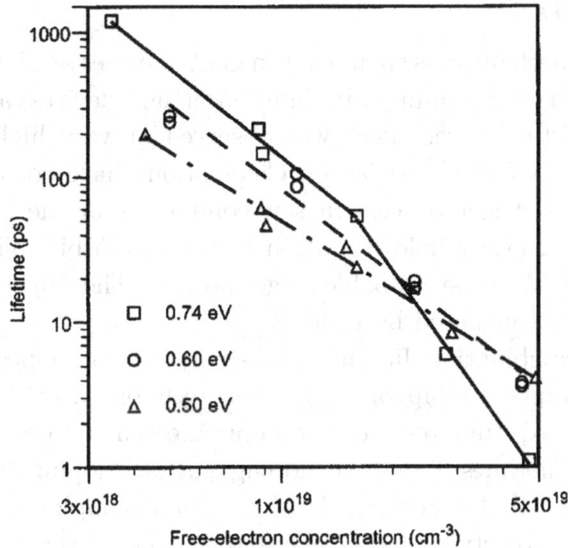

Fig. 8.3. Minority carrier lifetime on heavily doped In(x)Ga($1-x$)As with the three bandgaps of 0.74, 0.60, 0.50 eV. The lattice matched compositon (0.74 eV) show a steeper slope for doping levels above about 2×10^{19} cm^{-3}. [Reprinted from Ahrenkiel, R. K., Ellingson, R., Johnston, S., and Wanlass, M. (1998) *Appl. Phys. Lett.* **72**, 3470, with permission of AIP Publishing.]

Table 8.1. Auger Coefficients for the Alloy series In(x)Ga($1-x$)As. Listed are the fits to the data of Fig. 8.3 by Metzger *et al.*[12] and the data of Takeshima.[14]

In(x)Ga($1-x$)As Auger Coefficients				
Composition	Alloy	Band Gap	Auger	Reference
x	In(x)Ga($1-x$)As	(Eg) (eV)	$1/\tau = C_n^* n^x$	
0.53	In(.53)Ga(.47)As	0.74	$1.24\times10^{-29}n^{2.04}$	Metzger
0.66	In(.66)Ga(.34)As	0.60	$1.22\times10^{-27}n^{1.95}$	Metzger
0.78	In(.78)Ga(.22)As	0.50	$1.62\times10^{-28}n^{1.58}$	Metzger
1.00	InAs	0.35	$6.62\times10^{-27}n^{2}$	Takeshima

at the higher doping levels for the lattice matched composition. An explanation given by Huag[13] is a transition from phonon assisted Auger (n^{-2})doping dependence. For second order processes, $n^{-7/3}$ doping dependence is predicted.

8.2.3. *GaAs*

The Auger mechanism is significant in GaAs devices at high-injection levels,[15] that are common in light emitting devices and lasers. Evidence of the Auger effect was observed at very high injection levels in GaAs LEDs and lasers. Calculations have considered the details of the Auger process in semiconductors of the zinc blende structure.[16] A heavy hole transition to the spin-orbit split-off band appears to be the most probable Auger process. The Auger coefficient for GaAs was calculated by Haug.[17]

The recombination lifetime in heavily carbon-doped GaAs is very importance to bipolar transistor technology. Ahrenkiel and coworkers used time-resolved photoluminescence to measure the very short lifetimes found at doping carbon doping levels from $5 \times 10^{18}\,\mathrm{cm}^{-3}$ to $1 \times 10^{20}\,\mathrm{cm}^{-3}$. The lifetime samples were isotype double-heterostructures, Al(0.4)Ga(0.6)As/GaAs/Al(0.4)Ga(0.6)As grown by MBE.

The lifetime measurements were performed by up-conversion of time-resolved photoluminescence (TRPL). The photoluminescence (PL) photons were mixed with the pump-laser photons in a nonlinear crystal. The TRPL measurement system[18] has been described in Chapter 7.

Figure 8.4 shows TRPL data from two heavily doped samples, $4.9 \times 10^{19}\,\mathrm{cm}^{-3}$ and $1.05 \times 10^{20}\,\mathrm{cm}^{-3}$. The data were taken in the NREL laboratories and show low scatter as seen in the figure. Figure 8.5 is a plot of the Auger coefficient versus carrier concentration in this series of carbon-doped DHs. The Auger coefficient increases with as about $p^{0.97}$ (or $N_{\mathrm{A}})^{0.97}$ as seen in the data fit. The reasons for the concentration dependence include formation of an impurity band and other effects as discussed in the reference.

The concentration dependence of Auger recombination in carbon-doped GaAs can be written:

$$\frac{1}{\tau_{\mathrm{A}}} = 6.7 \times 10^{-48} p^{2.97}. \tag{8.13}$$

Using this expression as a general value for the Auger effect in GaAs, one finds the crossover (radiative to Auger domination transition) occurs at about $5 \times 10^{18}\,\mathrm{cm}^{-3}$.

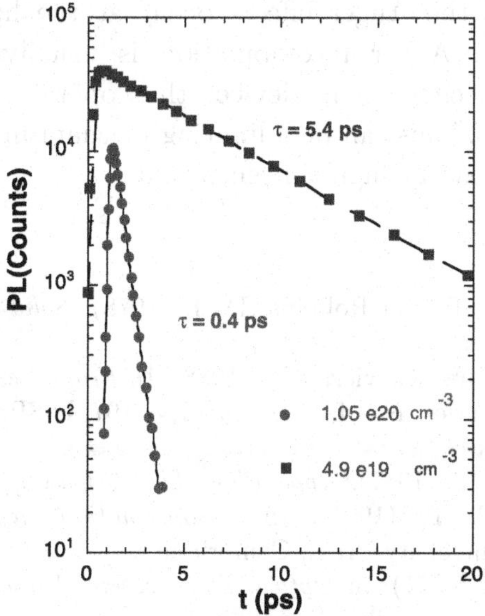

Fig. 8.4. The TRPL data on two carbon doped GaAs films with the indicated doping levels.

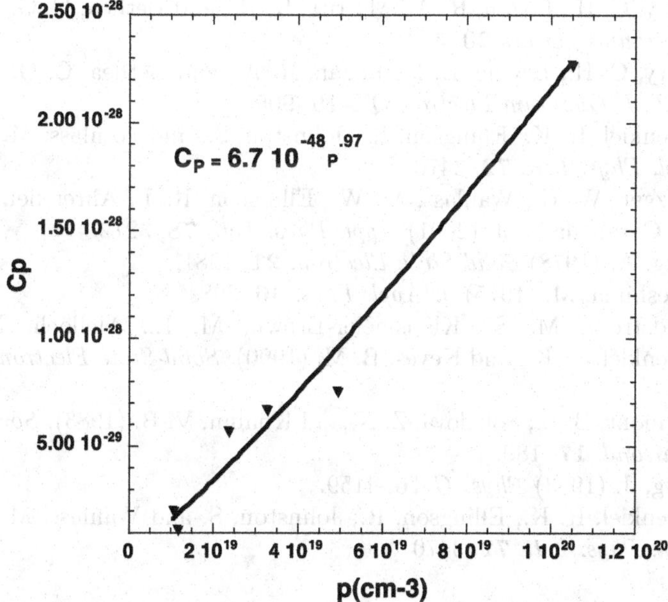

Fig. 8.5. The Auger coefficient on Carbon doped GaAs determined to the TRPL data by fitting five doping levels.

In summary, the Auger effects occur at the highest injection or doping levels. Auger recombination is usually the dominant recombination mechanism in devices that operate at high doping or injection levels. This can be a limiting mechanism in photovoltaic devices that operate at high concentration.

References

1. Landsberg, P. T. and Robbins, D. J. (1978). *Solid-St. Electron.* **21**, 1289.
2. Willander, M. and Grivickas, V. (1988). In *Properties of Silicon EMIS Datareviews* Series, No. 4, Ch. 8 pp. 195–197. INSPEC, IEE, London and New York.
3. Landsberg, P. T. (1987). *Appl. Phys. Lett.* **50**, 745.
4. Landsberg, P. T. (1991). In *Recombination in Semiconductors*, Cambridge University Press, Cambridge.
5. Pankove, J. I. (1971). In *Optical Processes in Semiconductors*, Dover Publications, New York.
6. J. Dziewior and W. Schmid, (1977) *Appl. Phys. Lett.* **31**, 346–348.
7. S. K. Pang and A. Rohatgi, (1993) *J. Appl. Phys.* **74**, 5554.
8. Sugimura, A. (1983). *Appl. Phys. Lett.* **42**, 17.
9. Henry, C. H., Logan, R. A., Merritt, R. R., and Bethea, C. G. (1984). *Electronics Letters* **20**, 359.
10. Henry, C. H., Levine, B. F., Logan, R. A., and Bethea, C. G. (1982). *IEEE J. Quantum Electron* **QE-19**, 905.
11. Ahrenkiel, R. K., Ellingson, R., Johnston, S., and Wanlass, M. (1998) *Appl. Phys. Lett.* **72**, 3470.
12. Metzger, W. K., Wanlass, M. W., Ellingson, R. J., Ahrenkiel, R. K., and Carapella, J. J. (2001) *Appl. Phys. Lett.* **78**, 3272.
13. Huag, A. (1978) *Solid-State Electron.* **21**, 1281.
14. Takeshima, M. (1975) *J. Appl. Phys.* **46**, 308s.
15. Lundstrom, M. S., Klausmeier-Brown, M. E., Melloch, M. R., Ahrenkiel, R. K., and Keyes, B. M. (1990). *Solid-State Electronics* **33**, 693.
16. Gel'mont, B. L., Sokolova, Z. N.,and Khalfin, V. B. (1983). *Sov. Phys. Semicond.* **17**, 180.
17. Haug, J. (1983) *Phys. C* **16**, 4159.
18. Ahrenkiel, R. K., Ellingson, R., Johnston, S. and Wanlass, M. (1998) *Appl. Phys. Lett.* **72**, 3470.

CHAPTER 9

Trapping Spectroscopy

9.1. Transient Recombination at Deep Defect Levels

In Chapter 3, the Shockley-Read-Hall theory was developed describing the interaction of point defects, with free carriers. Point defects are generally categorized as shallow or deep defects depending on the energy separation from the band of origin of the carrier.

These two kinds of localized states are generically called traps. We divide these two species to distinguish between traps and recombination centers. Traps are defined as those states that are energetically close to the band of origination of the carrier and the carrier can be thermally re-emitted to that band with a high probability. An excellent discussion of traps and trapping phenomena can be found in the classic textbook by A. Rose.[1] For example, electron traps are those defects with energy levels within a few KT of the conduction band. Likewise, hole traps are associated with defect levels that are a few KT from the valence band and have a finite capture cross section for holes. Recombination centers or deep states are positioned many KT from either band. When minority carries are captured in deep states, there is a much higher probably of majority carrier capture than thermal excitation to the band of origin.

We will analyze localized trap levels having density N_t and energy E_i relative to the valence band. Note that isolated states

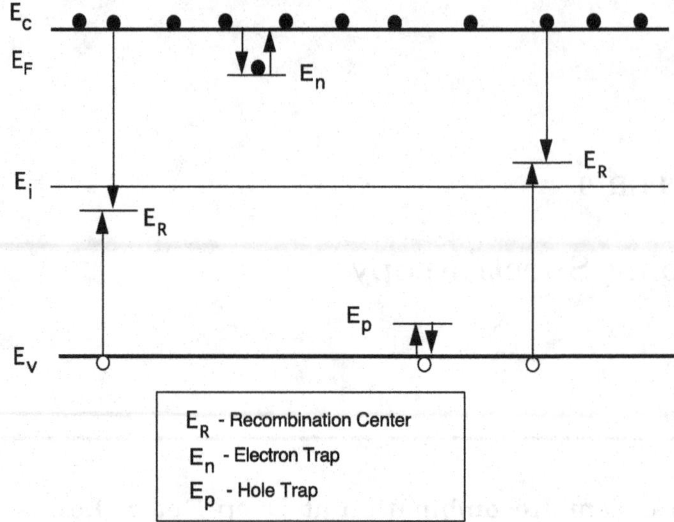

Fig. 9.1. Schematic representation of impurity levels that represent deep recombination levels and shallow electron and hole traps.

such as these will not form a continuous (impurity) band without wavefunction overlap between centers. We will analyze transitions involving exchange of carriers between the trap states and the CB and VB states.

9.1.1. *Carrier Capture and Emission by an Impurity*

Expressions for the transition rates can then be written based on the free carrier concentrations at each impurity center. Using the Fermi distribution function for the occupation of the trap level, we have calculated the rates of electron capture from the CB into an empty trap state. One describes hole capture as the annihilation of a hole by the emission of an electron from a filled trap state into an empty VB state.

One can calculate the equilibrium capture and emission rates by solving the rate equations. In Chapter 3, we described the electron capture rate, R_c, at an impurity center with energy level E_{T}:

$$R_c = V_{\mathrm{th}}\sigma_n n N_{\mathrm{T}}(1 - f(E_{\mathrm{T}})). \tag{9.1}$$

The emission rate, R_e, from this level is described by:

$$R_e = e_n N_{\mathrm{T}} f(E_{\mathrm{T}}). \tag{9.2}$$

At equilibrium, $R_c = R_e$ and the emission rate is calculated in terms of trap parameters.

$$e_n = V_{\mathrm{th}} \sigma_n n_i \exp\left(\frac{E_{\mathrm{T}} - E_i}{KT}\right), \tag{9.3}$$

where E_i the intrinsic energy level for the material (and very near midband). When, $(E_{\mathrm{T}} - E_\iota)/KT \gg 1$, the level is described as a trap or shallow trap as the emission rate becomes quite large. In that case, emission becomes more probable than hole capture by the occupied impurity center.

By analogy with the above derivation for electrons, the hole emission rate is:

$$e_p = V_{\mathrm{th}} \sigma_p n_i \left(\frac{E_i - E_{\mathrm{T}}}{KT}\right). \tag{9.4}$$

9.1.1.1. *Deep Level Recombination*

For deep levels: i.e. those near midgap, the probably of majority carrier capture generally dominates over the emission process. This process has been described in the literature for many years as the Shockley-Read-Hall (SRH) recombination process. One can write an steady state rate equation for all four possible processes show in the figure. For a steady state to occur, the net rate of electron capture must equal the net rate of hole capture.

$$R_c^n - R_e^n = R_c^p - R_e^p. \tag{9.5}$$

Here, the superscript designates the carrier type and the subscript designates the process, capture or emission. The SRH equation was previously derived in terms of this detailed balance between capture and emission. When the trap level is near the intrinsic level, E_i, the temperature dependent terms can be neglected as the intrinsic level n_i is very small compared to common doping levels.

When the defect energy level lies close to the conduction band, the hole-emission rate from the center is negligibly small. Thus, the defect center is defined as a trap or recombination center according to the applicable relationship:

9.1.2. *Trapping/Recombination Designation*

$$\begin{aligned} \text{Trapping:} \quad & e_n \gg R_c^p; \\ \text{Recombination:} \quad & R_c^p \gg e_n; \end{aligned} \tag{9.6}$$

The book by Rose gives a detailed analysis about the distinction between traps and recombination centers.

9.1.3. *SRH Equation for midgap Levels*

$$U_{\text{SRH}} = \frac{(np - n_i^2)}{\tau_n p + \tau_p n}. \tag{9.7}$$

The SRH equation is the standard model for deep level recombination effects in semiconductors. The equation shows that the recombination rate is a function of injection level. Materials such as silicon, that are generally dominated by defect recombination, show a nonlinear recombination behavior in transitions from low to high injection levels. When we calculate the recombination rate at any injection into a doped *p*-type material of doping level, N_{A}, we use the following previously used notations.

$$\begin{aligned} p &= N_{\text{A}} + \Delta p; \\ n &= n_0 + \Delta n; \\ n_0 N_{\text{A}} &= n_i^2. \end{aligned} \tag{9.8}$$

$$U_{\text{SRH}} = \frac{(n_0 + \Delta n)(N_{\text{A}} + \Delta p) - n_i^2}{\tau_n (N_{\text{A}} + \Delta p) + \tau_p (n_0 + \Delta n)}. \tag{9.9}$$

With electron-hole pair injection, $\Delta n = \Delta p$, and $np_0 = n_i^2$. The U_{SRH} rate can be written:

$$U_{\text{SRH}} = \frac{n_0 \Delta p + N_{\text{A}} \Delta n + \Delta n^2}{\tau_n (N_{\text{A}} + \Delta p) + \tau_p (n_0 + \Delta n)}.$$

For low injection levels, one makes the approximation:

$$\Delta n, \Delta p < N_{\mathrm{A}};$$

$$also,$$

$$n_0 = \frac{n_i^2}{N_{\mathrm{A}}} \ll \Delta n, \Delta p \tag{9.10}$$

$$U_{\mathrm{SRH}} = \frac{\Delta n}{\tau_n}; \quad p\text{-}type$$

$$U_{\mathrm{SRH}} = \frac{\Delta p}{\tau_p}; \quad n\text{-}type. \tag{9.11}$$

$$\begin{aligned} &\Delta n, \Delta p < N_{\mathrm{A}}; \\ &also, \\ &n_0 = \frac{n_i^2}{N_{\mathrm{A}}} \ll \Delta n, \Delta p. \end{aligned} \tag{9.12}$$

$$\begin{aligned} \tau_{\mathrm{low}} &= \tau_p(n\text{-}type), \\ \tau_{\mathrm{low}} &= \tau_n(p\text{-}type). \end{aligned} \tag{9.13}$$

And the excess carrier decay rate is described by the following expression in p-type semiconductors.

$$\Delta n(t) = \Delta n_0 \exp(-\sigma_n V_{\mathrm{th}} N_{\mathrm{T}} t). \tag{9.14}$$

However, as high injection is approached, the instantaneous lifetime can be derived as:

$$U_{\mathrm{SRH}} = \frac{N_{\mathrm{A}} \Delta n + \Delta n^2}{\tau_n(N_{\mathrm{A}} + \Delta n) + \tau_p \Delta p}; \tag{9.15}$$

Therefore, the high injection lifetime limit, for either n-type or p-type is:

$\Delta n \gg N_{\mathrm{A}};$

$$U_{\mathrm{SRH}} \simeq \frac{\Delta n}{\tau_n + \tau_p}; \tag{9.16}$$

The ratio of the high injection to low injection lifetime can also be written in terms of capture cross section.

$$R^{n}_{\text{inj}} \simeq \frac{1/\sigma_n + 1/\sigma_p}{1/\sigma_p} = \frac{\sigma_p + \sigma_n}{\sigma_n};$$
$$R^{p}_{\text{inj}} \simeq \frac{\tau_n + \tau_p}{\tau_n} == \frac{\sigma_p + \sigma_n}{\sigma_p}. \tag{9.17}$$

Each point defect has a specific value corresponding to the high-injection/low-injection lifetime ratio. Measurement of that value are used for the identification of the dominant defect in many cases.

9.1.4. *Injection Level Spectroscopy*

The ratio of high injection to low-injection lifetime has a unique value for each particular point defect and may be identified by one of the lifetime techniques.

This technique[2] has been called "Lifetime Spectroscopy" or injection level spectroscopy[3] by various authors. A number of techniques have been utilized to measure that ratio and identify the dominant deep level recombination center. Here, we have preferred to use time-resolved photoluminescence and time-resolved photoconductivity.

The capture cross sections of a given point defect for either electrons or holes changes markedly after the capture of either quasi-particle.

The capture process results in the addition of a positive or negative electronic charge state at the defect. For example, if the defect has a net charge of $+q$ before the capture of an electron, the defect becomes a neutral defect after electron capture. Consequently, the capture cross section for an additional electron decreases by orders of magnitude after the initial electron capture as the defect becomes neutral. When the $+q$ defects become saturated with electrons, recombination is dominated by the capture of holes at the neutral defect. Consequently, after very high injection pulses, one observes an initial large increase in the decay time until the traps return to the $+q$ state by emission of the captured electron to the conduction band.

In the case of p-type doping, the electron traps are quickly filled and the decay time is dominated by τ_p that may be much larger. For example, if the electron trap is positively charge (+1), the filled trap is neutral. Clearly $\sigma_n > \sigma_p$ in this case, and the high injection lifetime is much larger than the low injection lifetime.

9.1.5. *Transient Measurements of SRH Lifetime*

Because of these effects, a two component decay is commonly seen for SRH-dominated materials after a high injection pulse of carriers. The low-injection SRH lifetime is usually approximated by either τ_p or τ_n in n-type or p-type semiconductors, respectively. The recombination rate at the SRH level is controlled by the capture of the minority carrier by the defect. Thus the product of capture cross section σ and the defect density N_t controls the recombination rate. In low injection, the majority-carrier density N is much greater than the minority-carrier density ρ. After minority-carrier capture, the center is reset by the subsequent capture of a majority carrier. Because of the high concentration of majority carriers, capture occurs at a very high rate in the case of p-type doping. Thus in low injection, the SRH lifetime is determined solely by the minority-carrier capture rate. At high injection or with large cross-section ratios, this assumption cannot be made as will be discussed below.

A model calculation uses numerical integration to solve Eq. (9.18) for parameters that are representative of silicon wafers used in commercial photovoltaic devices.

$$\frac{dn}{dt} = \frac{dp}{dt} = -\frac{(np - n_i^2)}{\tau_n(p + n_i \exp(E_i - E_r)/KT) + \tau_p(n + n_i \exp(E_r - E_i)/KT)}. \tag{9.18}$$

In the calculation that follows, we assumed a defect energy level position at midgap ($E_r = E_i$). We assumed that the material is doped p-type to a level of $1 \times 10^{16}\,\text{cm}^{-3}$ and contains a single defect concentration of $1 \times 10^{13}\,\text{cm}^{-3}$. The following were used for the electronic parameters of the defect.

$$\sigma_n = 1 \times 10^{-14}\,\text{cm}^{-2};$$

$$\sigma_p = 1 \times 10^{-16}\,\text{cm}^{-2}$$

The calculated carrier lifetimes are;

$$\tau_n = 10\,\mu\text{s} \quad \text{and} \quad \tau_p = 100\,\mu\text{s}.$$

Curve A shows decay dominated by electron capture only. Curve B shows that the midgap level is partially filled with electrons, that are capturing holes at the majority carrier lifetime. Curve C shows complete saturation of the defect for about 25 μs, followed by a transition to the 10 μs lifetime that represents the electron lifetime.

Data similar to the calculated Fig. 9.2 are usually seen in photoconductive decay data of silicon wafers when high-injection is

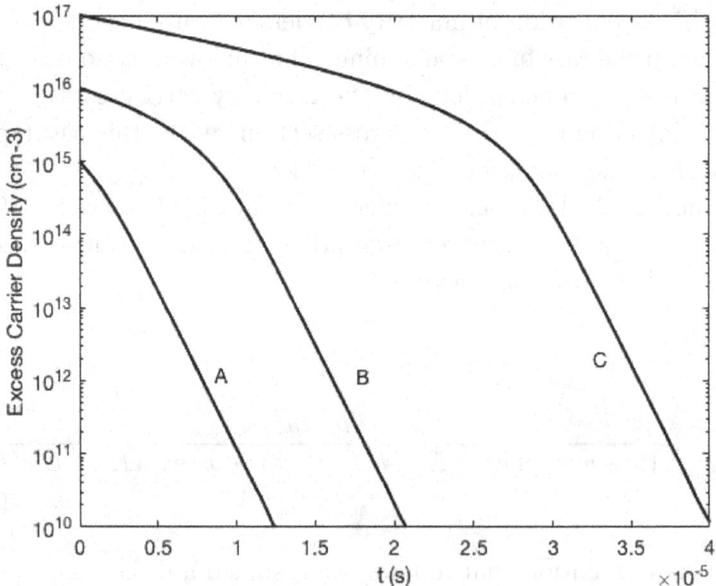

Fig. 9.2. Calculated excess carrier decay of a calculated model with a single midgap recombination center. The minority carrier cross section is ten time that of the majority carrier capture cross section. The calculated response to injection levels of A: $1 \times 10^{15}\,\text{cm}^{-3}$, B: $1 \times 10^{16}\,\text{cm}^{-3}$, and C: $1 \times 10^{17}\,\text{cm}^{-3}$.

applied. The ratio of majority carrier to minority-carrier lifetime is sometimes applicable to be unique to a specific defect.

The saturation of deep level defect has been seen by both PCD and TRPL.

One of the first observations of injection level dependence was observed[4] in the TRPL of $Al_{0.08}Ga_{0.92}$ As with a thickness of 5 μm and a doping level of 1×10^{16} cm^{-3}. A confinement layer is composed of $Al_{0.9}As_{0.1}$As grown by MOCVD. It is known that Al_xGa_{1-x}As, is direct bandgap semiconductor ($x < 0.5$). The TRPL was measured at three very different initial injection levels as shown on the figure. Consequently, these data show the SRH injection level dependence. The data of Fig. 9.3 of curve A shows an exponential decay (curve A) at the lowest injection level, but a transition to high injection-low injection transition at the highest injection level (curve C) with a lifetime ratio of about 2.5. The initial decay shows a rapid drop because of self-absorption.

For photoconductive decay, one measures $\Delta\sigma$ which is the product of $\Delta n^*\mu_n$ and $\Delta p^*\mu_p$. The variation of mobility with excess

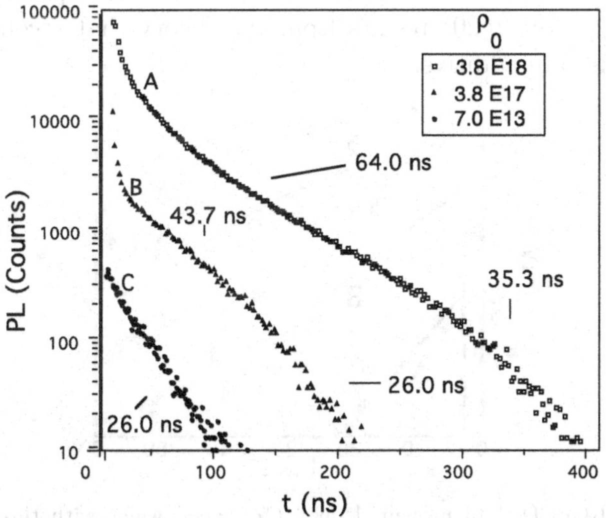

Fig. 9.3. The TRPL data[4] of thin-film $Al_{0.08}Ga_{0.92}$ As grown by MOCVD. The data shows self-absorption near $t = 0$ and the saturation of deep levels at high injection. [Reprinted from Ahrenkiel, R. K., Keyes, B. M., and Dunlavy, D. J. (1991). *J. Appl. Phys.* **70**, 225, with permission of AIP Publishing.]

carrier density, was discussed in Chapter 3, and may distort the decay from the analysis discussed above.

$$\frac{d\Delta\sigma(t)}{dt} = q\mu\frac{d\Delta(n)}{dt} + \Delta n\frac{d\mu}{dt}. \tag{9.19}$$

As $\mu(\Delta n)$ may exhibit significant changes in mobility (as shown in Chapter 3), a different impurity ratio may emerge from PCD data when compared to TRPL.

An example of photoconductive decay data is shown in Fig. 9.4. These data show similar characteristics to the model calculation of Fig. 9.2.

9.1.6. *Multiple Defect Centers*

When more than one deep-level defect type is present, the total SRH low-injection lifetime is described by a single minority-carrier lifetime τ where:

$$\frac{1}{\tau} = \sum_i \frac{1}{\tau_i} \tag{9.20}$$

Here τ_i is the Shockley-Read-Hall lifetime for each specific type of defect. Equation (9.20) results from the theory that recombination probabilities are additive.

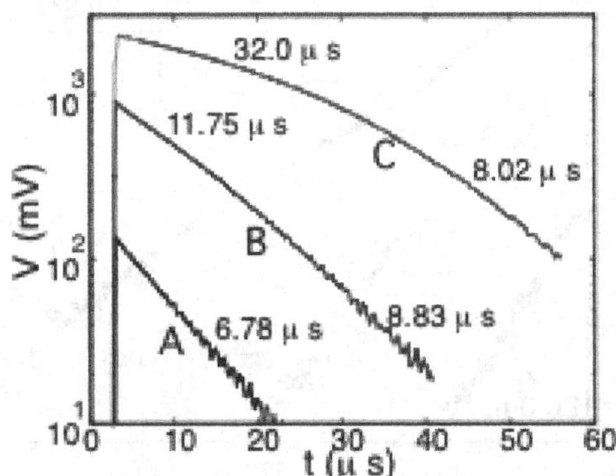

Fig. 9.4. RCPCD data measured on a CZ silicon wafer with three different pulsed injection levels. The injection levels for curves A, B, and C are 5×10^{13}, 5×10^{14}, and 5×10^{15} cm^{-3}, respectively. These data show the saturation of deep levels at the highest injection level.

9.2. Shallow Trapping: Theory

The term "trap" has been found in the literature for many decades. In the classic paper by Hornbeck and Haynes,[5] the authors suggest that minority carriers in silicon and germanium are temporarily "imprisoned" at special sites or imperfections. The authors contrast to what they describe as "deathnium" centers where the carrier was lost forever. Today, the common terminology is shallow traps and deep traps or recombination centers.

A deep trap here is currently defined as a defect level that captures an electron or hole. After capture, the carrier may recombine with a majority carrier and the carrier energy and transferred to lattice heat via phonons. This process is commonly called deep level or SRH recombination. If the carrier is thermally emitted to the band of origin, prior to any significant recombination, the behavior is called shallow trapping. This process is shown in Fig. 9.1. The trap is characterized with a capture cross-section, σ_t, and a trap density, N_t.

Because of the temperature terms in the SRH equation, a designated trap at a higher temperature may become a recombination center at low temperature.

One defies the shallow trap capture probability as:

$$\frac{1}{\tau_t} = V_{\text{th}}\sigma_t N_t. \tag{9.21}$$

Here, τ_t is the trap capture lifetime and V_{th} is the thermal velocity. The trap will re-emit the carrier at a rate e_n, with an emission time τ_e. Using the optical excitation function $G(t)$, one can write the rate equation for excess electrons as:

$$\frac{d\Delta n}{dt} = G(t) - \frac{\Delta n}{\tau_{\text{R}}} - V_{\text{th}}\sigma_t(N_t - \Delta n_t)\Delta n + \frac{\Delta n_t}{\tau_e}. \tag{9.22}$$

Here Δn is the density of free electrons and Δn_t is the density of trapped electrons. Also, τ_{R} is the recombination lifetime. We can write the companion rate equation for trap occupation as:

$$\frac{d\Delta n_t}{dt} = V_{\text{th}}\sigma_t(N_t - \Delta n_t)\Delta n - \frac{\Delta n_t}{\tau_e}. \tag{9.23}$$

The optical pulse also generates an equal number of holes, Δp. The holes are assumed to not interact with the trap, and therefore the rate equation for hole recombination is:

$$\frac{d\Delta p}{dt} = G(t) - \frac{\Delta p}{\tau_{\mathrm{R}}}. \tag{9.24}$$

Charge neutrality requires that:

$$\Delta n + \Delta n_t = \Delta p. \tag{9.25}$$

At steady state equilibrium, the trap capture and emission rates are equal (9.5), and therefore, the steady state excess free-electron and trapped electron densities are:

$$\begin{aligned} \Delta n_s &= G\tau_{\mathrm{R}}; \\ \Delta n_{t0} &= \frac{\Delta n_s}{\frac{\Delta n_s}{N_t} + \frac{\tau_t}{\tau_e}} \end{aligned} \tag{9.26}$$

Here Δn_s and Δn_{t0} are the free and trapped electron steady-state densities, respectively. In the steady state, when the $G\tau$ product is larger than the trap density, N_t, the traps are essentially filled and $\Delta n_t \sim N_t$.

For transient measurements, we first look at Eq. (9.22) and apply very low injection conditions, $\Delta n \ll N_t$, for an impulse injection pulse at $t = 0$. Then, the approximation to the differential rate equation is:

$$\frac{d\Delta n}{dt} = -\frac{\Delta n}{\tau_{\mathrm{R}}} - \frac{\Delta n}{\tau_t}. \tag{9.27}$$

Thus, under these injection conditions, the excess carriers decay as:

$$\Delta n = \Delta n_0 \exp\left(\frac{-t}{\tau_{\mathrm{R}}}\right). \tag{9.28}$$

Thus, the very low injection lifetime is a combination of recombination and the trap capture rates. Under higher injection condition, when $\Delta n < N_t$, then $\Delta n_t \sim N_t$ and the solution to Eq. (9.22) is approximately:

$$\Delta n = \Delta n_0 \exp\left(\frac{-t}{\tau_\mathrm{R}}\right). \tag{9.29}$$

Thus, at injection levels that "fill" the trapping centers, the transient decay reflects the recombination lifetime.

At times much longer than τ_R, trap emission dominates the excess carrier density. In this time regime and $\tau_e \gg \tau_\mathrm{R}$, the time-dependent trap density comes from the solution of Eq. 3 with $\Delta n \ll \Delta n_t$:

$$\Delta n_t = \Delta nt_0 \exp(-t/\tau_e). \tag{9.30}$$

During transient photoconductivity measurements, the density of excess majority carriers, Δp, is nearly equal to the Δn_t, the density of trapped minority carriers. Therefore, a long photoconductive response is observed (as τ_e is usually much larger than τ_R), and is given by:

$$\Delta\sigma(t) = q\mu_p \Delta p(t) \sim q\mu_p \Delta n_{t0} \exp(-t/\tau_e). \tag{9.31}$$

Thus, the signature of shallow traps is a long photoconductive "tail" that is observed in the photoconductive transient.

Figure 9.5 shows the RCPCD response of two films using the tripled pulse YAG with excitation wavelength of 355 nm. Film A was subjected to the standard $CdCl_2$-annealing process, whereas Film B was as-grown. These films show the long photoconductive-decay behavior that is attributed to trap emission response. The excess carrier density shows a long-term decay time of about 9 μs. Time-resolved shows that the recombination lifetime (at $t \sim 0$) is in the nanosecond range for these films. The annealed Film A shows an enhanced response near $t = 0$ with an initial lifetime of 0.29 μs. The behavior is indicative of trap filling and the initial response includes both recombination and trap-filling behavior. Film B has a weaker response that may be indicative of only trap capture and emission. The effect of $CdCl_2$ treatment on the response of these films is likely to be increased grain size and weaker grain boundary recombination.

To verify the shallow trapping explanation of these data, the RCPCD measurements were performed at temperatures between

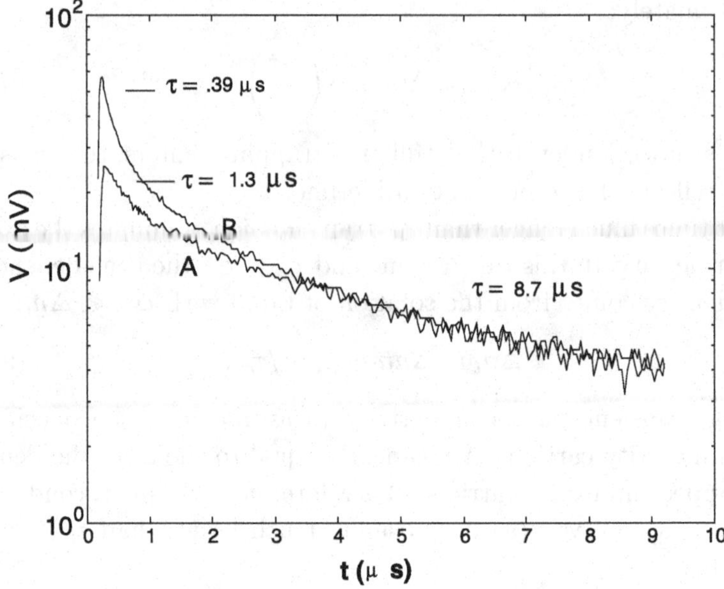

Fig. 9.5. RCPCD photoconductive decay data of a polycrystalline thin film of CdTe. Curve A: response of film after anneal in $CdCl_2$; curve B: response of film as-grown.

80 and 300 K in a cryogenic apparatus. Figure 9.5 shows the photoconductive decay of the annealed sample A over the temperature range of 175 to 296 K. There is a steep decrease in the long-term signal with decreasing temperature. The behavior is indicative of indicative of the decrease in emission rate with temperature, and a "freeze out" of the trap emission signal. The shallow trap transitions to a deep trap and becomes a recombination center at cryogenic temperatures. The latter behavior is observed in the much shorter initial lifetime, indicating the transition from trapping to recombination behavior.

For the steady-state case, one may easily show that the low- and high-injection photoconductivity coefficients are respectively,

$$\Delta\sigma_{\rm L} = qG\tau_{\rm R}[\mu_n + \mu_p(1 + \tau_e/\tau_t)] \tag{9.32}$$

$$\Delta\sigma_{\rm L} = qG\tau_{\rm R}[\mu_n + \mu_p]. \tag{9.33}$$

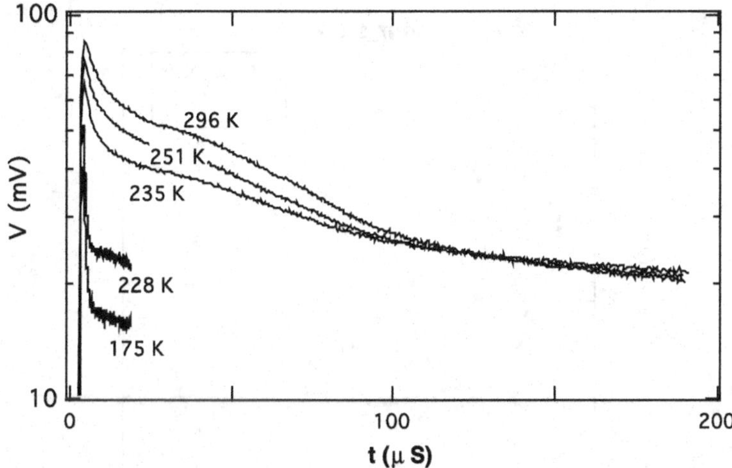

Fig. 9.6. The photoconductive response of the annealed CdTe film of Figure 9.6 as a function of sample temperature.

These equations will be applied later to multi-crystalline silicon that is measured by steady state photoconductivity techniques. For the steady-state case, one may easily show that the low- and high-injection photoconductivity coefficients are respectively,

$$\Delta\sigma_{\mathrm{L}} = qG\tau_{\mathrm{R}}[\mu_n + \mu_p(1 + \tau_e/\tau_t)] \tag{9.34}$$

$$\Delta\sigma_{\mathrm{L}} = qG\tau_{\mathrm{R}}[\mu_n + \mu_p]. \tag{9.35}$$

These equations describe the data measured on multi-crystalline silicon. Figure 9.7 shows data on a multicrystalline wafer using the quasi-steady state photoconductivity (QSSPC) technique. The wafer was masked into four separate circular areas and the data measured separately on each area.

The data show an increased photoconductive lifetime at low injection originating from trap capture/emission. As the injection level is increases, the traps are filled and the data reflect the recombination lifetime.

One of the common measurement errors in analyzing semiconductors is attributing trap emission signals to recombination lifetime.

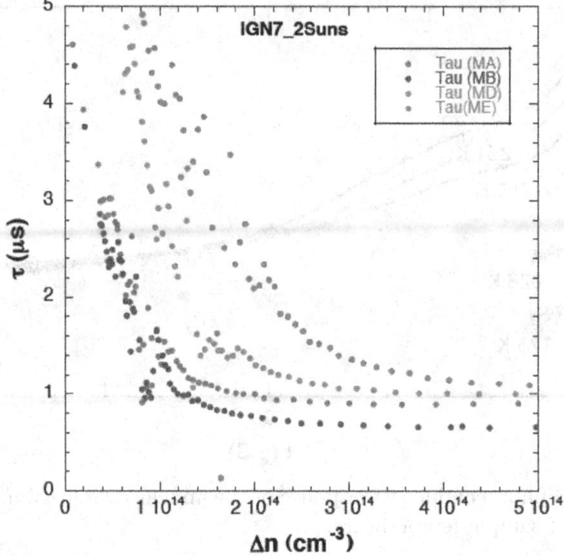

Fig. 9.7. Trapping effects measured by QSSPC of four masked areas of the same multicrystalline silicon wafer shown in Fig. 6.6.

The measurement needs to use injection level and/or temperature or other methods to avoid making this common mistake in data analysis.

9.2.1. *Variable Temperature PCD*

The RCPCD technique is particularly adaptable to cryogenic measurement. S. W. Johnston[6] has reported a number of variable temperature measurements that enable the identification of the dominant recombination or trapping mechanism. For example, when radiative recombination is dominant, the lifetime decreases with decreasing temperature at the B-coefficient increases as $T^{-1.5}$. Johnston and coworkers measured the PC lifetime of epitaxial In(0.53)Ga(0.47)As DHs that is lattice-matched to InP. This particular sample was Zn-doped p-type to a level of $1 \times 10^{16}\,\text{cm}^{-3}$. The RCPCD of this sample is shown in Figure 9.8 below and the decay time increases to very large values near liquid nitrogen temperature. By plotting the PC decay data and applying the emission rate equation (9.4), the data fit produced a trapping level at $E_v + 0.16\,\text{eV}$. This level

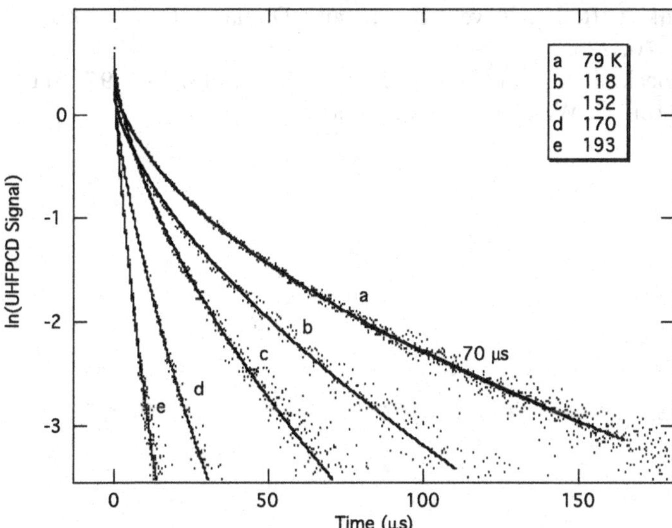

Fig. 9.8. The RCPCD decay of an epitaxial In(0.53)Ga(0.47)As thin film as a function of temperature. The photoconductive decay was produced by a 1064 nm YAG laser with several ns pulse width. Analysis of these data indicate a hole trap at $E_v + 0.16\,\text{eV}$.

coincides with a known-substitutional Fe level in the forbidden gap of In(0.53)Ga(0.47)As. The radiative lifetime at 79 K is about 0.1 μs using the B-coefficient discussed earlier. Clearly, the lifetime changes with temperature shown here indicate delayed recombination caused by multiple trapping. The 70-μs-lifetime seen at the lowest temperature is then the inverse of a trap emission rate at that temperature. Distinguishing between trapping and recombination becomes a challenging task in the interpretation of data. Temperature variation measurements are one of the more definitive methods for identifying the underlying physical processes.

References

1. Rose, A. (1963). *Concepts in Photoconductivity and Allied Problems*, John Wiley and Sons, New York.
2. Rein, S. (2004). *Lifetime Spectroscopy*, Springer.
3. Ahrenkiel, R. K. and Johnston, S. W. (2000). *Surface Engineering*, **16**, 54.

4. Ahrenkiel, R. K., Keyes, B. M., and Dunlavy, D. J. (1991). *J. Appl. Phys.* **70**, 225.
5. Hornbeck, J. A. and Haynes, J. R. (1955). *Phys. Rev.* **97**, 311.
6. Johnston, S. W. (private communication).

CHAPTER 10

Steady State Techniques

Steady state optical illumination produces excess carriers that may be detected by sensors of photoconductivity or photoluminescence. Here we will calculate the steady state excess carrier density as a function of basic parameters such as lifetime, diffusion length, and mobility. These calculations can then serve as a basis to interpret specific measurements. This calculation will be basic to all steady state measurement techniques of excess carrier density. Contained in the excess carrier density versus incident intensity are parameters of bulk lifetime and surface recombination velocity.

10.1. Semi-Infinite Structures

For a steady state, monochromatic optical input, the diffusion equation is applied to a semi-infinite p-type bulk material with surface recombination velocity S. The solution is:

$$\rho(x) = \frac{\alpha\tau I_0}{1-(\alpha L_n)^2}\exp(-\alpha x) + A\exp(-x/L_n) \tag{10.1}$$

The constant of integration A is determined from the boundary condition at the front surface:

$$D\left[\frac{d\rho(x)}{dx}\right]_{x=0} = S\rho(0). \tag{10.2}$$

Here, S is the surface recombination velocity in units of cm/s.

The solution for the constant A is:

$$A = -I_0\tau\frac{\alpha}{1-\alpha^2L^2}\frac{\alpha L + SL/D}{1+SL/D}. \tag{10.3}$$

The complete solution for excess minority-carrier density as a function of position is:

$$\rho(x) = I_0\tau\frac{\alpha}{1-\alpha^2L_n^2}\left[\exp(-\alpha x) - \frac{\alpha L_n + SL_n/D}{1+SL_n/D}\exp(-x/L_n)\right]. \tag{10.4}$$

Here, α is the absorption coefficient of the monochromatic light, L_n and D_n are the electron diffusion length and diffusion coefficient, respectively.

The total excess minority-carrier density is obtained by integrating Eq. (x) over all space or from zero to infinity here.

$$\rho_t = \int_0^\infty \rho(x)dx = I_0\tau\left(\frac{\alpha L}{1-\alpha^2L^2}\right)\left(\frac{1}{\alpha L} - \frac{\alpha L + SL/D}{1+SL/D}\right). \tag{10.5}$$

We see that the excess carrier density is equal to $I_0 * \tau$ times a function of absorption coefficient α, S, L, and D. In the limit of $S = 0$, we see that:

$$\rho_t = I_0\tau. \tag{10.6}$$

However, there are very large reductions in the steady state concentrations when the surface recombination velocity is sufficiently large, i.e., larger than 100 cm/s according to forthcoming calculations. The surface recombination affects both steady state photoconductivity and steady state photoluminescence measurements. Examples will be shown in data that are obtained from steady state measurements.

We will discuss the various measurement techniques that are commonly used to measure the excess carrier density, ρ_t, using the steady state methods. Such measurements are commonly used to extract values of the carrier lifetimes and other properties.

10.2. Quantum Efficiency

Figure 10.1 shows a plot of Eq. (10.5) is applied to single crystal silicon with a range of parameters that are typical of commercially available material. We will call this calculation the minority-carrier excitation spectra (MCE) and the incident wavelength has been varied from 300 nm to 900 nm. The silicon absorption coefficients of Green[1] were used for the calculation. The primary variable in this calculation is the surface recombination velocity S. The parameter for SRV or S has been allowed to vary from 1 to 10^4 cm/s. This is a range

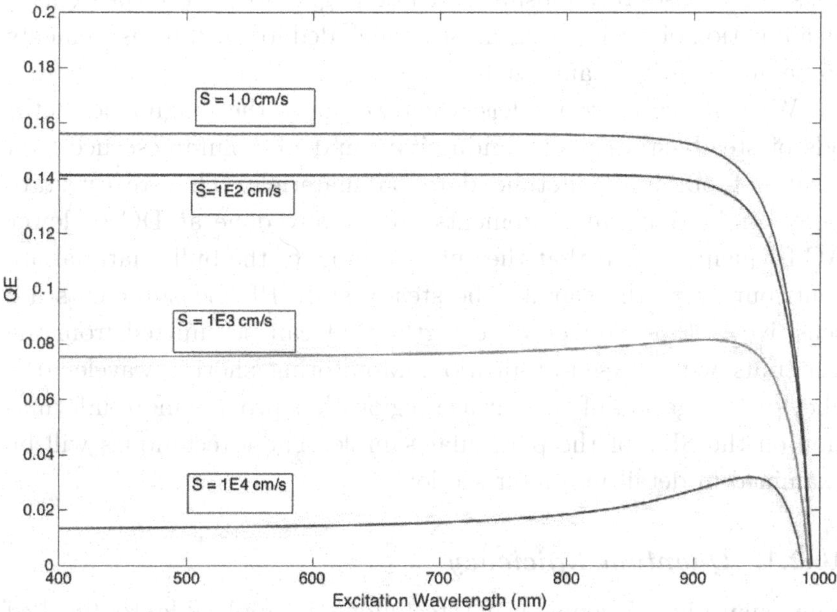

Fig. 10.1. The calculated steady state minority carrier density versus excitation wavelength with the SRV as a parameter.

typical of passivated and unpassivated silicon wafers. Each curve has been normalized to 1.0 as a maximum value. The calculation shows little sensitivity to S values less than 100 cm/s. However, there are large drops in the absolution value between $S = 1 \times 10^3$ and $S = 1 \times 10^4$ (about a factor of 3 for the former to 17 for the latter values of S). The latter are more typical of unpassivated silicon, and must be accounted for in any steady state measurement that are dedicated to estimating a lifetime value.

Figure 10.1 shows a calculation of the relative minority carrier density versus excitation wavelength for crystalline silicon using a bulk lifetime of 100 μs. Lifetime values of this magnitude or larger are found in crystalline silicon grown by the float zone technique or advanced Czochralski (CZ) technique. We see from the figure that the surface recombination produces large reductions in the QE response and controls minority-carrier density. For S values increasing from 100 to 1000 cm/s, the equilibrium minority-carrier density decreases by almost an order of magnitude. The surface effect, as a function of wavelength, must be included for such measurements to be quantitatively analyzed.

We will use these models when we apply the diagnostic methods of steady-state photoconductivity and photoluminescence as a means of obtaining lifetime data on materials. The steady state photoconductivity measurements are usually done at DC or lower AC frequencies such that the entire volume of the bulk material are contributing to the signal. The steady state PL measurements are sensitive to those longer wavelengths that can be emitted from the materials without self-absorption. Monitoring shorter wavelengths allow interrogation of the surface region that provide more information on the SRV of the particular sample. These techniques will be examined in detail in a later section.

10.2.1. *Quantum Efficiency*

The quantum efficiency (QE) technique is applicable to finished devices and is widely used by the photovoltaics community. The basic measurement involves measuring the short-circuit current of a device as an incident monochromatic beam is incident on the front surface of the device. A monochromatic light beam is usually obtained by a dispersing white light source and scanning the output wavelength over some, finite wavelength range. The ratio of the generated electrons to the incident light flux (in photons $s^{-1}\,cm^{-2}$) at that wavelength is called the external quantum efficiency (EQE). This includes light reflected from the incident surface and other losses. The ratio of the generated electrons (or holes) to the absorbed photons is the internal quantum efficiency (IQE). A schematic of a typical $n+/p$ silicon diode that is used in the photovoltaic technology is shown in Fig. 10.2. We will do an analysis of the quantum efficiency of this device over the range of visible wavelengths that are absorbed by silicon. The calculations will solve for the short-circuit current using the diffusion equation over the visible and near infrared absorption spectrum of silicon.

W_{E} is the length of the n^+ emitter and W_{B} the length of the p-type base. The alignment of the emitter and base Fermi levels results in band bending and depletion layer formation at the interface. The

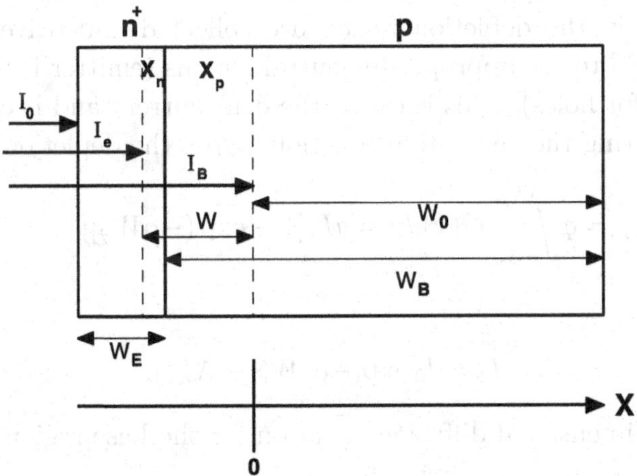

Fig. 10.2. The device structure that will be used in the forthcoming calculations.

depletion width is the emitter is given as X_n and the base width is X_p such that $X_n + X_p = W$, the total depletion width. The total depletion width is related to the n and p doping levels as:

$$W = \sqrt{\frac{2\varepsilon_s V_{bi}}{q}\left(\frac{1}{N_A} + \frac{1}{N_D}\right)};$$

$$where:$$

$$V_{bi} = \frac{KT}{q}\ln\left(\frac{N_A N_D}{n_i^2}\right). \qquad (10.7)$$

The optical generation function incident on the device is:

$$G(x) = \alpha I_0 \exp(-\alpha x); \qquad (10.8)$$

The monochromatic light intensity reaching the depletion boundary in the emitter is described by the attenuated optical generation function:

$$I_e = I_0(1 - R)\exp[-\alpha(W_e - X_n)]. \qquad (10.9)$$

where R is the front surface reflection and α is the absorption coefficient at the specific wavelength. We will assume that all carriers generated in the depletion region are collected and driven by the electric field to the appropriate neutral regions (emitter for electrons and base for holes). This is called the drift current and is calculated by integrating the generation function across the depletion regions.

$$J_{dr} = q\int_{-X_n}^{X_p} G(x)dx = qI_e\left[1 - \exp\left(-\alpha W\right)\right]; \qquad (10.10)$$

where,

$$I_e = I_0 \exp(-\alpha(W_E - X_n)).$$

The one-dimensional diffusion equation for the base region is:

$$D_n \frac{d^2\Delta n}{dx^2} - \frac{\Delta n}{\tau_n} = G(X) = I_B \exp(-\alpha x); \qquad (10.11)$$

where,

$$I_{\mathrm{B}} = I_0 \exp[-\alpha(W_{\mathrm{E}} + X_p)]. \tag{10.12}$$

The diffusion equation for electrons in the base region is:

$$\Delta n(x) = A\cosh(x/L_n) + B\sinh(x/L_p) + C\exp(-\alpha x); \tag{10.13}$$

Here C is the trial particular solution for this generation function. A and B are constants to be determined from the boundary conditions. By substituting the third term into the diffusion equation, we find the constant C to be:

$$C = \frac{\alpha I_{\mathrm{B}}\tau_n}{\alpha^2 L_n^2 - 1}. \tag{10.14}$$

The boundary conditions are:

$$\Delta n(0) = 0;$$
$$qD_n \frac{d\Delta n(W_0)}{dx} = -qS_n \Delta n(W_0). \tag{10.15}$$

The first condition results from the short-circuiting of the junction. The second condition results from the flow of diffusion current into the surface states at the back contact. The first condition provides the constant A and gives $A = -C$. The excess electron density in the base is then:

$$\Delta n(x) = \frac{\alpha I_{\mathrm{B}}\tau_n}{\alpha^2 L_n^2 - 1}[\exp(-\alpha x) - \cosh(x/L_n)] + B\sinh(x/L_n). \tag{10.16}$$

After some consolidation of terms, one finds the constant B from the second (back surface) boundary condition.

$$B = \frac{\alpha I_{\mathrm{B}}\tau_n}{\alpha^2 L_n^2 - 1} \times \frac{\sinh(W_0/L_n) - Z_n\cosh(W_0/L_n) + \exp(-\alpha W_0)(Z_n + \alpha L_n)}{cosh(W_0/L_n) + Z_n\sinh(W_0/L_n)}. \tag{10.17}$$

Then the electron population in the base is:

$$\Delta n(x) = \frac{\alpha I_{\mathrm{B}} \tau_n}{\alpha^2 L_n^2 - 1} \times \left(\begin{array}{l} [\exp(-\alpha x) - \cosh(x/L_n)] \\ + \left[\dfrac{\begin{array}{c} \sinh(W_0/L_n) - Z_n \cosh(W_0/L_n) \\ + \exp(-\alpha W_0)(Z_n + \alpha L_n) \end{array}}{cosh(W_0/L_n) + Z_n \sinh(W_0/L_n)} \right] \sinh(x/L_n) \end{array} \right) \tag{10.18}$$

Here Z_n is a dimensionless parameter proportional to the back surface SRV, and I_B was defined above.

$$Z_n \equiv \frac{S_n L_n}{D_n}, \tag{10.19}$$

current flowing from the short circuit current comes from the addition of the drift current generated in the depletion region and the diffusion base into the depletion region. The diffusion current into the *pn* junction is:

$$J_{\mathrm{SC}}^{diff} = qD_n \left. \frac{d\Delta n(x)}{dx} \right]_{x=0} = \frac{-\alpha^2 q L_n^2 I_{\mathrm{B}}}{\alpha^2 L_n^2 - 1} + \frac{qD_n}{L_n} B. \tag{10.20}$$

The total diffusion currents are combined and written as:

$$J_{\mathrm{SC}}^{diff} = \frac{q\alpha L_n}{\alpha^2 L_n^2 - 1} I_{\mathrm{E}} \exp(-\alpha W) \times \left(\frac{\begin{array}{c} \sinh(W_0/L_n) - Z_{\mathrm{N}} \cosh(W_0/L_n) \\ + \exp(-\alpha W_0)(Z_n + \alpha L_n) \end{array}}{\cosh(W_0/L_n) + Z_n \sinh(W_0/L_n)} - \alpha L_n \right). \tag{10.21}$$

The drift current from the junction, space-charge region is:

$$J_{\mathrm{SC}}^{dr} = qI_{\mathrm{E}} (1 - \exp(-\alpha W)). \tag{10.22}$$

Here, the input optical intensity at the point labeled I_{E} is:

$$I_{\mathrm{E}} = (1 - R)I_0 \exp\left(-\alpha(W_{\mathrm{E}} - X_n)\right). \tag{10.23}$$

The total short-circuit current is given by adding the two components.

$$\mathrm{IQE} = J_{sc}/(1 - R)^{*}I_0; \tag{10.24}$$

In finished photovoltaic devices, the front surface reflection R is reduced to nearly zero by anti-reflection coatings.

This model is used to calculate the internal quantum efficiency of a silicon wafer devices is plotted with carrier lifetime as a parameter. The wavelength is varied from 300 nm to 950 nm, using the silicon absorption data of Green. The lifetime values are shown in the figure. The four curves represent calculations with recombination lifetimes of 1,10,100, and 1000 μs. These are a range of lifetimes found in as-grown multicrystalline silicon (1 μs) to float-zone or high quality Czychroski-grown silicon crystal silicon (1000 μs or more). The model neglected any recombination in the depletion layer, so the short wavelengths 450 nm or less, were collected with high efficiency. Here, the depletion width, w, is calculated as 0.32 μm and the absorption length or 450 nm

Illumination is 0.39 μm. Therefore, the carriers are primarily generated in the depletion region and accelerated to the majority carrier side of the junction. At long wavelengths, the lower absorption coefficient resulting from the indirect transitions, produce deep absorption in the base. The shorter lifetimes and diffusion lengths have a very large influence on current collection at the junction. Therefore, there is a significant drop in the IQE in the indirect absorption wavelength range.

The apparatus for measuring QE has involved a wavelength scanning light instrument using a monochromator that disperses a white light source. A detector measures the short-circuit current and a beam splitter measures a portion of the excitation energy at each wavelength. The ratio of current to beam intensity is calibrated to provide the QE as a function of wavelength. This arrangement has been a standard method of QE measurement. The limitation of the apparatus is that data collection is somewhat time consuming.

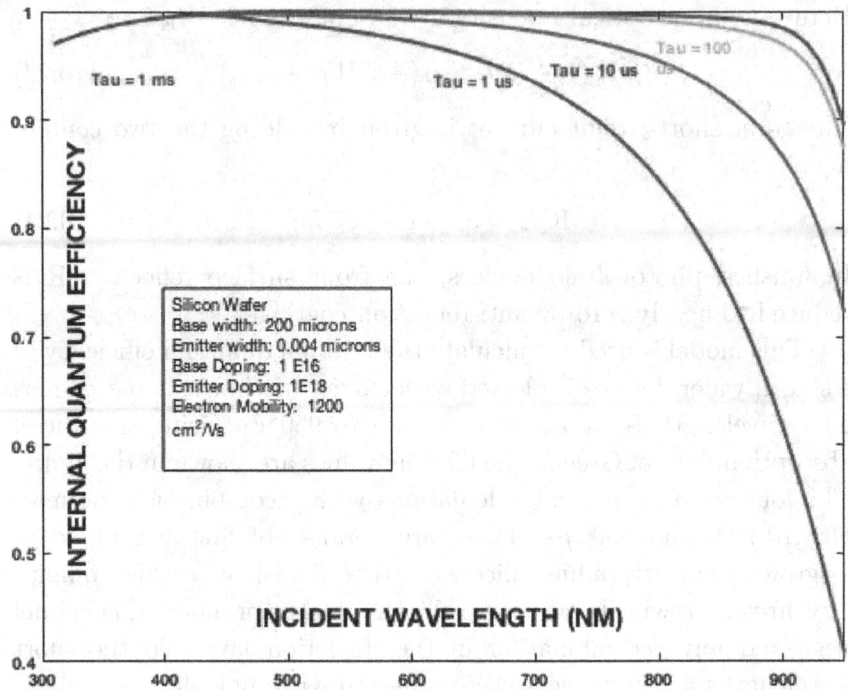

Fig. 10.3. Calculations of the internal quantum efficiency of a 200-μm-thick silicon wafer with the bulk carrier lifetime as the parameter of each curve.

A recent innovation allows for much more rapid data collection.[2] A real time full spectrum system provides the QE of a silicon solar cell in less than one second. A schematic diagram of that apparatus is shown in Fig. 10.4. An array of LEDs with a range of wavelengths that cover the range 300 nm to 1000 nm, is used as the light source. Each individual LED is modulated at a unique frequency in the kilohertz range. The photocurrents are superimposed in the output but retain the individual frequencies of the specific excitation source. These analog data are digitized and analyzed by a computer using a fast Fourier transform (FFT) algorithm.

The data of Fig. 10.5 are from an early measurement that uses ten LEDs to measure the QE a silicon photovoltaic diode. The QE from conventional scanning apparatus are shown by the dotted line. The conventional system required twenty minutes for data acquisition,

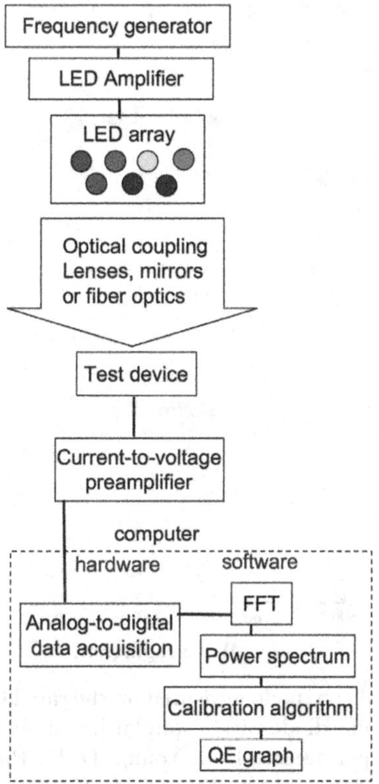

Fig. 10.4. Schematic of the apparatus used for rapid QE measurements. [© [2019] IEEE. Reprinted, with permission, from Young, D. L., Pinegar, S., Stradins, P., and Egaas, B. (2008). New real-time quantum efficiency measurement system, 33rd IEEE PVSC, San Diego, CA.]

whereas the RTQE system required about one second. A later version of the apparatus used 57 LEDs to give better wavelength resolution. This system is commercially available with an LED source that cover the wavelengths range 300 to 1580 nm.[3]

10.3. Steady State Photoluminescence Spectroscopy

Measurement of the photoluminescence (PL) spectra of semiconductors is a valuable component of semiconductor characterization. Historically, PL intensity has been used as a quick and nondestructive technique to measure the quality of a given material.

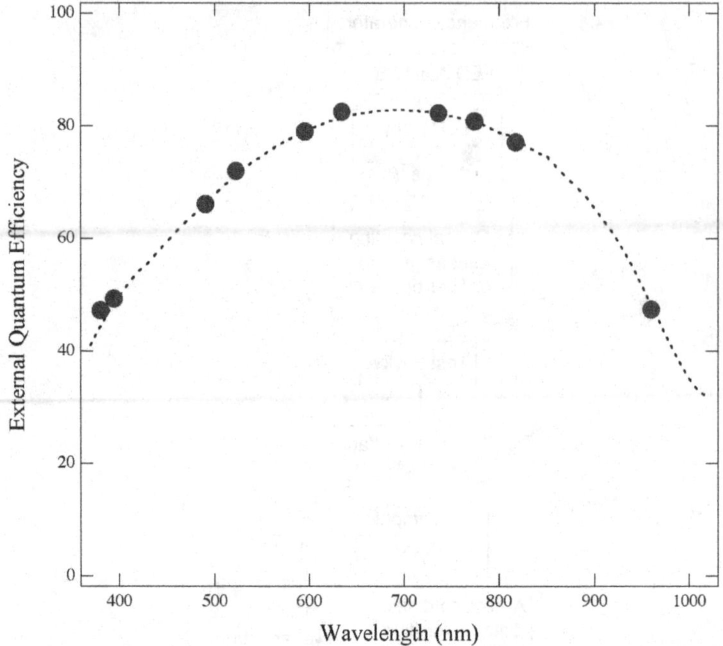

Fig. 10.5. Data on a silicon diode measured on the rapid QE system of Fig. 10.4. Data measured on the same diode with a standard, scanning QE system. [© [2019] IEEE. Reprinted, with permission, from Young, D. L., Pinegar, S., Stradins, P., and Egaas, B. (2008). New real-time quantum efficiency measurement system, 33rd IEEE PVSC, San Diego, CA.]

The band-to-band PL intensity is an indicator of recombination lifetime and quick method to find the relative material quality. One can also use PL to find the band-gap E_g of new material of unknown composition. The quantum efficiency is expressed the ratio of radiative recombination to the total recombination. This relationship can be written:

$$\eta = \frac{R_{rad}}{R_{total}} = \frac{R_{rad}}{R_{rad} + R_{nr}}. \tag{10.25}$$

Here, R_{rad} and R_{nr} correspond to radiative and nonradiative recombination rates. As the Recombination rates are inversely proportional to the respective lifetimes, the quantum efficiency can be written;

$$\eta = \frac{\tau_{n\mathrm{R}}}{\tau_{\mathrm{R}} + \tau_{n\mathrm{R}}}. \tag{10.26}$$

As the nonradiative lifetime becomes short compared to the radiative lifetime in silicon and other indirect materials, the efficiency approaches zero. The opposite is true for epitaxial GaAs and other high quality compound semiconductors. In these cases, η approaches unity. Measuring the PL efficiency of a device has become a very popular diagnostic for detecting nonradiative recombination mechanisms. More recently, PL mapping has become a powerful to method to diagnose the spatial profile of poly-crystalline silicon. In the latter, there is a wide variation in carrier lifetime because of extended defects (e.g. dislocations) and point defects. Digital cameras have been successfully used to make such maps of large polycrystalline areas.

In the case of silicon, one can write:

$$\eta = \frac{BN\tau_{n\mathrm{R}}}{1 + BN\tau_{n\mathrm{R}}} \approx BN\tau_{n\mathrm{R}} \tag{10.27}$$

as $\tau_{\mathrm{R}} \gg \tau_{n\mathrm{R}}$.

With a spatially uniform excitation source, I_0, the localized PL intensity measured by a digital camera is:

$$I_{\mathrm{PL}} = I_0(1 - R_s)^* BN\tau_{n\mathrm{R}}. \tag{10.28}$$

Here, R_s is the surface reflection and N is the doping concentration that may vary with position in multi-crystalline silicon. The nonradiative lifetime includes both bulk and surface recombination that is usually varies across the sample.

Figure 10.6 is steady state PL data measured on three wafers of silicon with variable lifetimes and doping densities. The PL spectra were measured while immersed in iodine methanol passivating solution[4] such that the surface recombination was reduced to a very small value. The doping density was measured by the capacitance-voltage technique, using a mercury probe to make a temporary metal-semiconductor diode. The surfaces of both samples were mechanical polished by the same procedure and etched in a dilute HF solution before immersion into the measurement cell. The minority carrier lifetime was measured by the RCPCD technique. The PL intensity ratio here is proportional to the doping density in accordance with

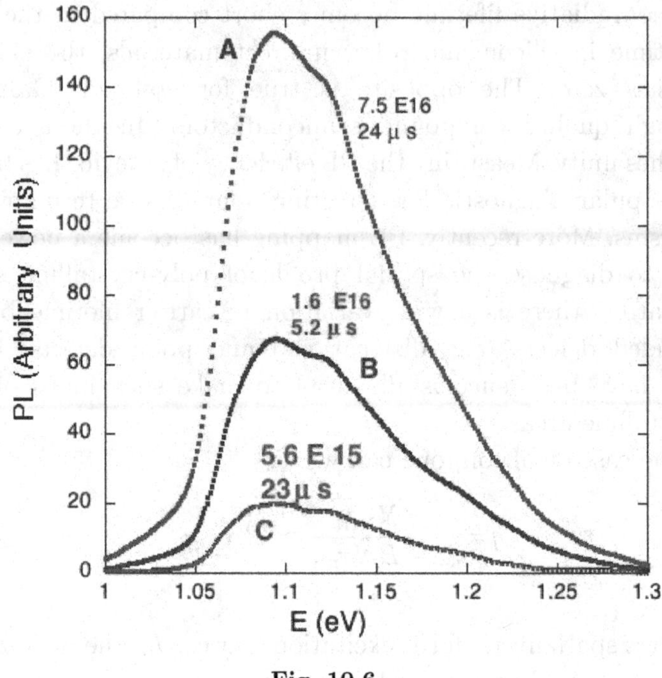

Fig. 10.6.

Eq. 10.28. For more accurate lifetime determination, one must know the local doping density to get a correlation to lifetime.

The data shows that the measured lifetime that samples A and C have about the same lifetime but the doping density is about 13 times larger in sample A. The PL signal in sample A is about 8 times larger than that of sample C. The primary variable in this case is the doping density. Other unknowns are surface reflectivity and measurement artifacts.

High resistivity wafers show that most significant effects of surface recombination as shown in Fig. 10.7. These data were obtained from PL measurements on a high resistivity float zone wafer, with a lifetime of about 1 ms. Curve A is the PL measurement on the bare wafer and curve B is the measurement in methanol/iodine solution after the standard HF etching process. The PL intensity is highly reduced by surface recombination as seen by comparing curves A and B.

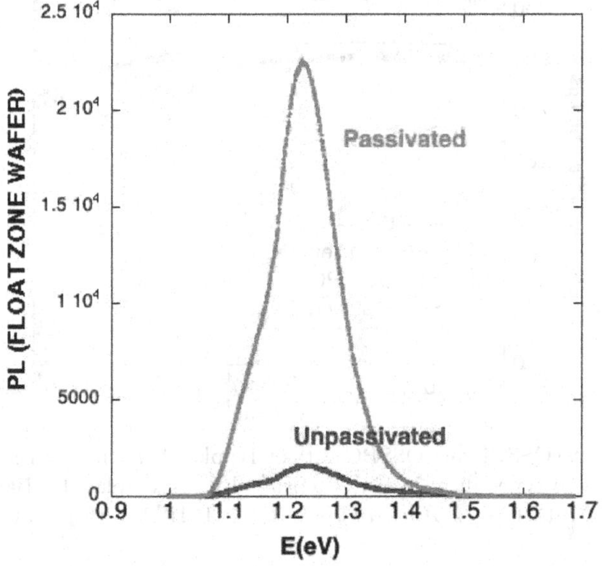

Fig. 10.7.

10.4. Quasi-Steady State Photoluminescence (QSSPL)

Trupke[5] and coworkers described a quasi-steady state photoluminescence (QSSPL) technique in 2005. This is a variation of the QSSPC technique that monitors the photoluminescence rather than the photoconductivity. The data can be analyzed by modifying equation

$$I_{\mathrm{PL}}(t) = B(p_0\Delta n(t) + \Delta n(t)^2).$$
$$\Delta n = G_0\tau; \tag{10.29}$$

where $\Delta n(t)$ is generated by a slowly varying light source. In the work described above, the light source was an 870 nm LED array with an output of 1.5 W and modulated at 15.5 Hz. The LED output was measured with a calibrated silicon sensor and the peak amplitude corresponded to a generation rate of $4.6 \times 10^{17}\ \mathrm{cm}^{-3}$. The PL signal was calibrated by measuring transient PCD and PL signals in parallel with the QSSPL signal. The data from the QSSPL, QSSPC, and transient measurement series are shown in Fig 10.8.

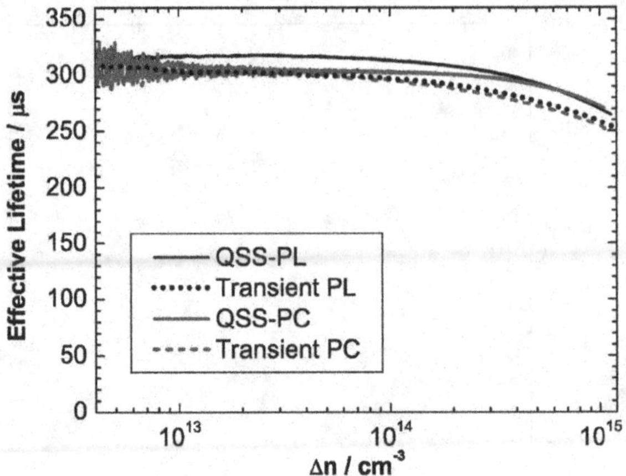

Fig. 10.8. The QSSPL and QSSPC data of Trupke *et al.* with transient methods overlaid on an *n*-type silicon wafer. [Reprinted from Trupke, T., Bardos, R. A. and Abbott, M. D. (2005). *Appl. Phys. Lett.* **87**, 184102, with permission of AIP Publishing.]

Figure 10.9 shows data using QSSPL and calibrated by QSSPC to provide a lifetime versus injection level measurement by Wang[6] and coworkers. The CZ wafer was passivated with silicon nitride. The data was measured as-grown and with three other post-growth procedure to assess the influence on carrier lifetime. The data show the small variations in lifetime that were attributed to the procedures.

One inherent advantage of the QSSPL method is that carrier mobility is not a factor in the output whereas mobility must be included the analysis of QSSPC. A number of publications have been produced that pertain to more refined calibration methods.

10.5. The Phase Shift Technique

The phase shift technique[7] has been used since the 1970s to measure the carrier lifetime in direct-band gap materials that have strong radiative photoluminescence. The standard technique used high repetition rate pulses (typically 50 to 100 MHz) from a later to produce pulses of luminescence decay from a thin III-V film. Using a fast photo-detector to capture the PL pulses, the phase shift between the excitation and PL pulses was proportional to the carrier lifetime. This system was used at NREL in the 1980s to measure the carrier

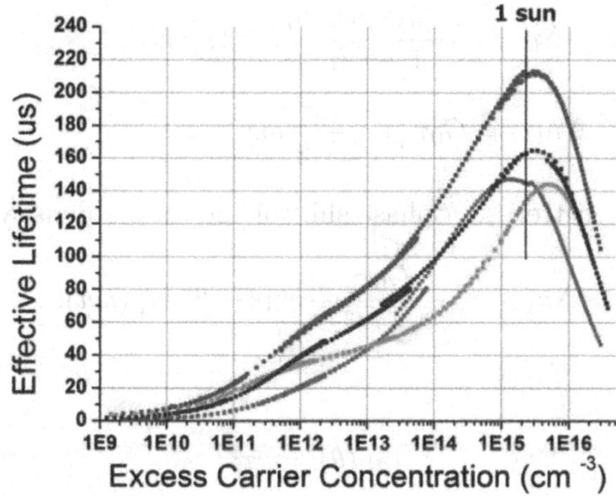

Fig. 10.9. The QSSPL measurements of a *p*-type, passivated wafer calibrated with complimentary QSSPC measurements by Wang and coworkers. [Reprinted from *Energy Procedia* **55**, Wang, X., Juhl, M., Abbott, M., Hameiri, Z., Yao, Y., and Lennon, A., Injection-dependent carrier lifetime analysis of recombination due to boron-oxygen complexes in wafers passivated with different dielectrics by QSSPL and QSSPC, 169–178 (2014) with permission from Elsevier.]

lifetime in thin films. The data analysis[8] was discussed in detail in an article by the first author (RKA) in the "Current Topic in Photovoltaics" series.

Using the continuity equation in a confinement structure with a sinusoidal optical input, the equation is written:

$$\frac{\partial \Delta n(t)}{\partial t} = G_0 \exp(i\omega t) - \frac{\Delta n(t)}{\tau}; \tag{10.30}$$

Here, G_0 is the amplitude of the sinusoidal light source.
Using a trial solution: where ϕ is an unknown phase shift.

$$\Delta n(t) = \Delta n_0 \exp(i\omega t - \varphi); \tag{10.31}$$

$$\frac{\partial n(t)}{\partial t} = i\omega \Delta n(t) = G_0 \exp(i\omega t) - \frac{\Delta n(t)}{\tau}; \tag{10.32}$$

Inserting the trial solution and cancelling the common time dependence term:

$$\Delta n_0 = \frac{G_0 \exp(i\varphi)}{\frac{1}{\tau} + i\omega}; \quad \text{or simplifying:} \tag{10.33}$$

$$\Delta n_0 = \frac{\tau G_0 \exp(i\omega\phi)}{1 + i\omega\tau}; \quad \text{Then;}$$

$$\Delta n(t) = G_0\tau \frac{1 - i\omega\tau}{1 + \omega^2\tau^2} \exp(i\omega t); \tag{10.34}$$

This can be written as a phase shift of the carrier density.

$$\Delta n(t) = \frac{G_0\tau}{\sqrt{1 + \omega^2\tau^2}} \exp(-i\theta) \exp(i\omega t); \tag{10.35}$$

Here:

$$\tan(\theta) = \omega\tau;$$

where τ is the carrier lifetime and ω is the frequency of the light source. The measurement of the phase shift between light source and the detected PL signal provides a lifetime value. For a PL signal, the modulated photoluminescence is expressed by:

$$I_{\mathrm{PL}}(t) = \frac{BNG_0\tau}{\sqrt{1 + \omega^2\tau^2}} \exp(-i\theta) \exp(i\omega t); \tag{10.36}$$

Here, B is the radiative coefficient of silicon and N is the doping level. This excludes the Δn^2 term of high injection. Thus, one gets an absolute lifetime value that is independent of the other parameters of the measurement. This can be performed as a function of injection level for a second parameter and give additional information.

This technique is applicable when the decay time, τ, is much shorter than the period of the excitation source, T, where: $T = \frac{2\pi}{\omega}$; For the mode locked laser excitation source used in the original configuration and repetition rates of 100 MHz, one is limited to lifetimes of less than 100 ns. This configuration is obviously inappropriate for silicon.

Giesecke[9] and coworkers used a 880 nm diode laser for excitation of silicon wafers grown by the float zone technique. Their modulation frequency varied from 1.0 to 300 Hz and the maximum incident intensity varied from 10^{16} to 10^{19} cm^{-2} s^{-1}. They were able to measure lifetimes as short as 20 μs.

10.6. Photoluminescence Mapping

Spatial mapping has been recently successfully applied to polycrystalline CdTe that is a key photovoltaic technology. In this case, $\tau_{n\mathrm{R}} > \tau_{\mathrm{R}}$ as this is a direct bandgap material. The QE then is approximately:

$$\eta = \frac{\tau_{n\mathrm{R}}}{\tau_{\mathrm{R}} + \tau_{n\mathrm{R}}} \approx 1 - \frac{\tau_{\mathrm{R}}}{\tau_{n\mathrm{R}}}. \tag{10.37}$$

Again, the PL map indicates the location of impurity clusters as the nonradiative processes become more prominent and the defect concentration decreases the total lifetime. Here, the PL map produces a signal according to:

$$I_{\mathrm{PL}} = I_0(1 - R)\left(1 - \frac{1}{\tau_{n\mathrm{R}} BN}\right). \tag{10.38}$$

Here the local doping density is also a factor. The PL intensity of each domain can be interpreted as being proportional to:

$$I_{\mathrm{PL}} \approx N * \tau. \tag{10.39}$$

where N is the localized doping density.

An application of the previous calculations will be shown to the recent technique of PL mapping of silicon. This has been shown to be an extremely useful technique to characterize multicrystalline silicon wafers that are the primary material being used in low-cost residential photovoltaic technology. The PL map shows grain boundaries and defective areas using low-cost digital cameras. The next section will derive the bulk lifetime of silicon as:

$$QE = \frac{I_{\mathrm{PL}}}{I_0(1 - R_s)} = BN\left(\frac{1}{\tau_{n\mathrm{R}}} + \frac{1}{\tau_{\mathrm{S}}}\right). \tag{10.40}$$

Here, $\tau_{n\mathrm{R}}$ is the bulk, nonradiative recombination and τ_{S} is the surface component. We are neglecting the bulk radiative contribution to intensity as it is much weaker in silicon. The previous calculations showed a more accurate influence of surface recombination on the excess carrier density as a function of excitation wavelength.

Trupke and coworkers[10] introduced the concept of PL mapping of multicrystalline silicon wafers in 2006. They used a 15 W-815 nm

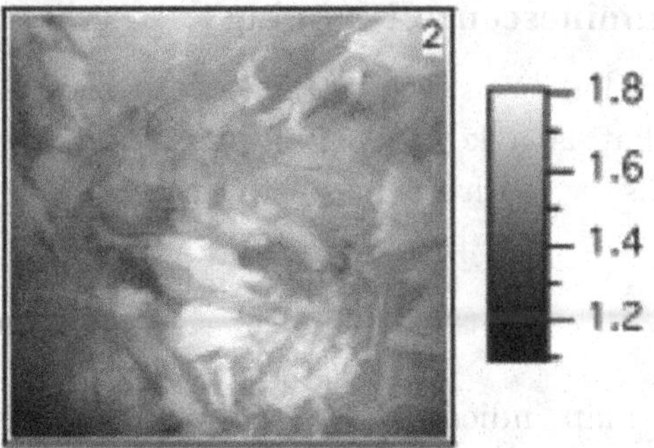

Fig. 10.10. PL image of a multicrystalline silicon using a 600 second exposure. The wafer is unprocessed with no surface passivation. Lifetime on the right is found by normalizing pixel count to a microwave lifetime map.

diode laser to illuminate wafers up to $8 \times 8\,\text{cm}^2$ and recorded the PL intensity with a megapixel, silicon CCD camera. The reflected excitation light was excluded with a long pass glass filter, such that only the wafer PL signal was recorded. Quality images of the microstructure were recorded and the variations in intensity were linked to lifetime variation.

The NREL group reported PL imaging on four, unpassivated multicrystalline Si wafers and one wafer passivated with a thermal oxide. The total lifetime was dominated by surface recombination, but information about the bulk material was learned from PL imaging. However, the unpassivated wafers required 600 seconds of exposure, whereas the passivated wafer required 1.0 exposure to get adequate signal to noise. The PL intensity was correlated with carrier lifetime but measuring the same areas with a commercial microwave mapping apparatus. The wafers were illuminated with 808 nm laser diode from the backside. A Princeton Instruments/Acton PIXIS 1024BR camera was used to record the PL map.

The surface passivation of the wafer increased the PL intensity by two orders of magnitude. The passivation resulted in reduction of the data collection time by the same factor.

One sees from these images that grain structure and low lifetime areas are visible in both passivated and unpassivated cases. PL mapping has become a very valuable characterization tool in the

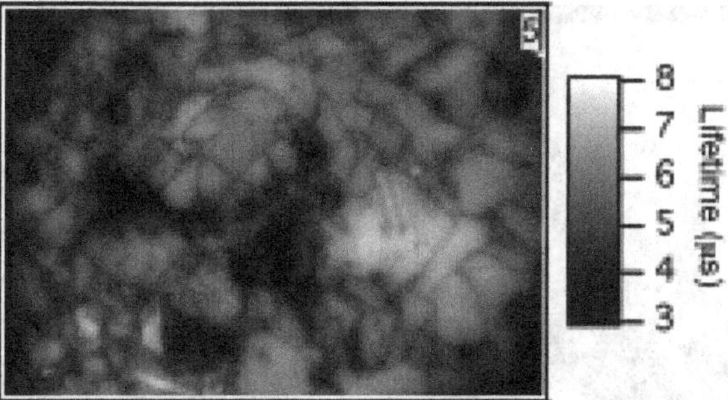

Fig. 10.11. PL image of a multicrystalline wafer using a 1.0 second exposure. The wafer is passivated with a thermal oxide. Lifetime on the right is found by normalizing pixel count.

multicrystalline solar cell industry. The PL intensities depend upon the sample's material quality, defects, surface passivation, and excitation intensity. PL images can be collected with exposure times ranging between 10 and 30 s for unpassivated wafers.

10.7. PL Mapping of Thin Films

Many of the thing film materials used in photovoltaics and other optoelectronic devices are made from direct bandgap materials. PL mapping has been applied to the characterization of these thin films in recent years. For a direct bandgap material, radiative recombination is the dominant mechanism for materials of device quality. The radiative efficiency can be written is this case (i.e. $\tau_{n\mathrm{R}} >= \tau_{\mathrm{R}}$) as;

$$\eta_{\mathrm{R}} = \frac{R_{\mathrm{R}}}{R_{\mathrm{R}} + R_{n\mathrm{R}}} \simeq 1 - \frac{R_{n\mathrm{R}}}{R_{\mathrm{R}}}. \tag{10.41}$$

For high quality epitaxial films of GaAs grown in the DH structure, the radiative efficiency is very close to 1.0. For polycrystalline thin films, such as those used in residential photovoltaics, the radiative efficiency varies widely and are effective indicators of defective regions. These areas include grain boundaries and areas of poor grain quality. S. W. Johnston and coworkers[11] have made extensive maps of thin film photovoltaic materials such as copper indium gallium selenide (CIGS) and cadmium telleride (CdTe).

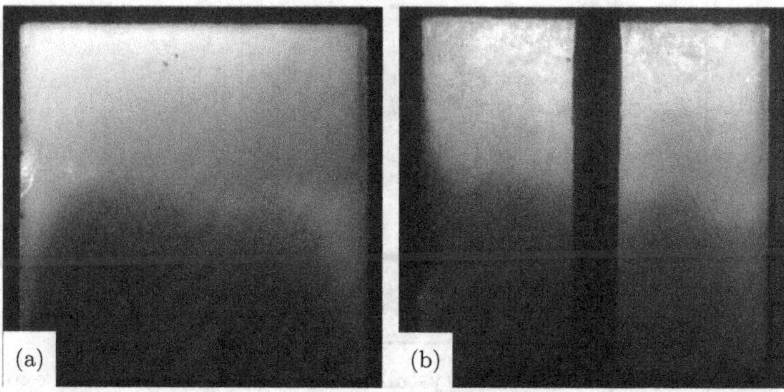

Fig. 10.12. (a) PL image of as-grown CIGS film. (b) PL image of CIGS film after CdS deposition. [Reprinted with permission from Johnston, S., Unold, T., Repins, I., Sundaramoorthy, R., Jones, K. M., To, B., Call, N., and Ahrenkiel, R. (2010). *J. Vac. Sci. Technol.* **A28**, 665. Copyright [2019], American Vacuum Society.]

The thin film material, CuInGaS (CIGS) is a direct bandgap material with a bandgap of about 1.0 eV. Figure 10.12 shows the PL imaging[12] of a CIGS film deposited on a molybdenum film coated glass substrate. The light source used for these images was a tandem of four 30 W, 810 laser diodes coupled to the sample with optical fibers. Engineered diffusers provide uniform excitation over the area of the active solar cell. The camera used for detection of the PL signal was a Princeton Instruments PIXIS 1024 BR camera (1024 × 1024 pixels) and a 50 mm macroimaging lens. Figure 10.12(a) is a map for an as-grown sample that shows the p-type base of the photovoltaic cell. The film was then coated with CdS by chemical bath deposition and the PL image shown Fig 10.12(b). One sees an increase in PL intensity after the CdS deposition, and the result correlates with the electrical activity of the device. The grain boundaries are too small to resolve here. The PL map intensity tracks the quality of the film using the radiative efficiency as a metric.

Figure 10.13 shows a PL map of a CdTe thin film as grown (Fig. 10.12(a)) and after $CdCl_2$ processing (Fig. 10.12(b)). The brightness increase is the typical response to the $CdCl_2$ and this property is correlated with a more efficient solar cell. The granular structure is very pronounced in CdTe films.

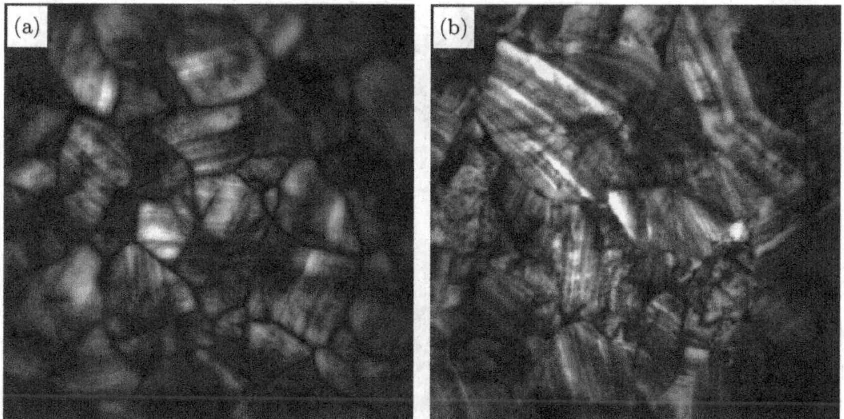

Fig. 10.13. PL maps of CdTe thin films. A. As-grown film: B-another region of film after $CdCl_2$ processing. PL brightness increase and grain growth are observed.

10.8. Electroluminescence Mapping

The PL mapping technique has proven to be very useful for analyzing the multi-crystalline silicon wafers that are currently the most widely deployed residential technology. One of the limitations of the PL mapping process is to acquire a light source that can uniformly illuminate large area devices or panels.

A recent innovation has been to use electroluminescence as an equivalent technique that is easily adaptable to large area devices or panels. When a solar diode is externally driven into forward bias, the device acts like a light emitting diode (LED) and emits band-to-band radiation at the bandgap. The intensity of that emission is a function of critical cell parameters, such as the carrier lifetime. For example, in the typical $n+/p$ emitter-base configuration of silicon PV cells, the electrons ejected into the base decay primarily by SRH recombination but there is competitive radiative recombination that is usually measureable by a light sensor or camera. The intensity of the emitted radiation is approximately given by the product of the generation factor and carrier lifetime.

S. W. Johnston[13] recently demonstrated this technique on a large area panel. These data are shown in Fig. 10.14 on the panel that is

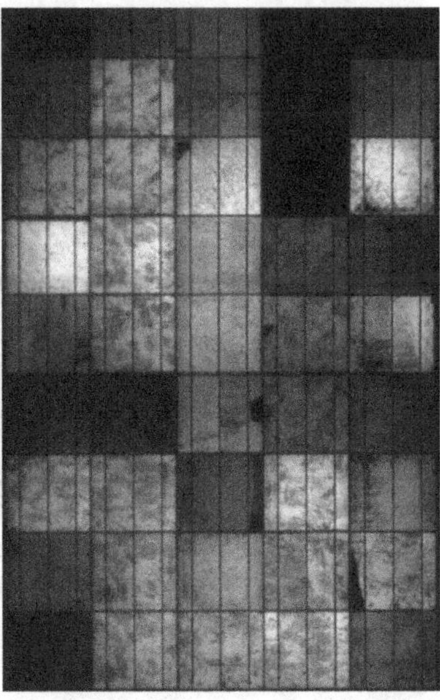

Fig. 10.14. EL image a PV module used to diagnose a degraded panel of 6 × 6 inch multi-crystalline silicon cells measured by S. W. Johnston. [© [2019] IEEE. Reprinted, with permission, from Johnston, S. W. (2015). 42nd IEEE Photovoltaics Specialists Conference, New Orleans, LA, pp. 1–6.]

driven into forward bias by an externally applied voltage source that is supplying 2.5 A of current. The EL image was taken by a Princeton Instruments 1024 BR charge coupled device camera. The exposure time here is 45 seconds.

The individual cells are 6 × 6 inches in size and display a large variation of EL intensities for various reasons. The variations are caused by shorts and lifetime degradation. The granular structure of the higher quality cells is apparent. Figure 10.15 compares the image of some of these same cells and with the PL signal of the same cell. This EL method offers the promise of doing on-site diagnosis of panels that are deployed in a rooftop or other application.

The limitation of EL is that a complete device is required to allow forward bias excitation of a signal. This technique is not applicable to bare wafers.

Johnston introduced the approach shown in Fig. 10.16 below. Rather than applying a voltage source to generate EL, this apparatus

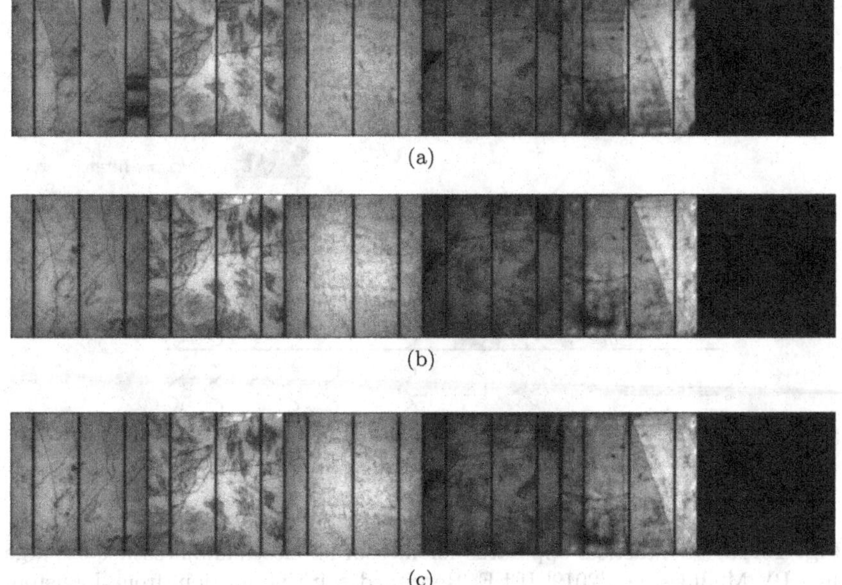

(a)

(b)

(c)

Fig. 10.15. (a) EL image of row #5 (from top) of the panel above. (b) PL image of the same row (#5) of cells. (c) EL signal was generated by an external light source. [[2019] IEEE. Reprinted, with permission, from Johnston, S. W. (2015). 42nd IEEE Photovoltaics Specialists Conference, New Orleans, LA, pp. 1–6.]

generates a signal by using an LED as a contactless voltage source. The open circuit voltage is generated on a remote area of the device and is propagated by the common contact to normally "dark" areas of the device.

Figure 10.15(a) shows the EL image of row #5 (of Fig. 10.14) generated with the external forward bias and Fig. 10.15(b) shows the PL image of the same row. These photos indicate that there are only minor differences between the two methods of excitation. Next, an EL signal was generated by the light source using the configuration of Fig. 10.16. These data from row #5 of the modules is shown in Fig. 10.15(c). The EL map of Fig. 10.15(c) is quite comparable with the maps of Figs. 10.15(a) and 10.15(b). In summary, all three methods of excitation provide the same basic information. One can choose the method that is most suitable to the measurement environment.

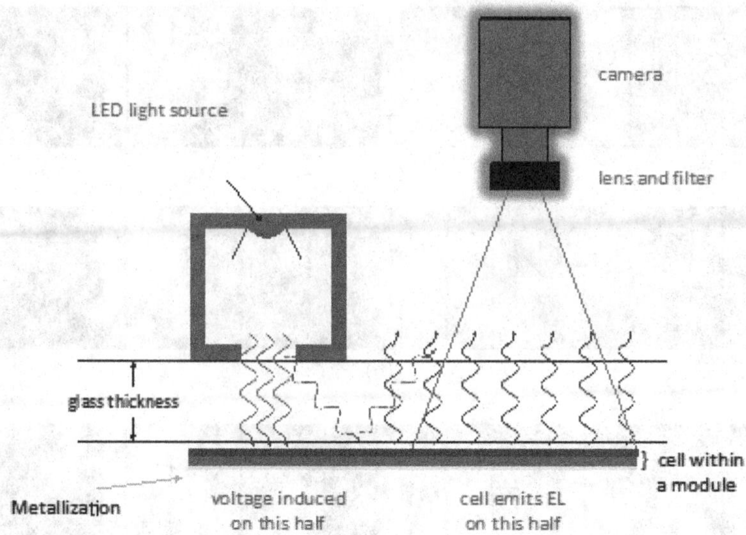

Fig. 10.16. Contactless apparatus for measuring electroluminescence on large area PV Modules. [© [2019] IEEE. Reprinted, with permission, from Johnston, S. W. (2015). 42nd IEEE Photovoltaics Specialists Conference, New Orleans, LA, pp. 1–6.]

Using a similar configuration to that shown in Fig. 10.6, Davis[14] and coworkers replaced the single wavelength LED with a variable wavelength array similar to that of Fig. 10.4. Each wavelength is modulated with a specific frequency and the induced EL at the bandgap is identified by a specific frequency that is linked to the excitation wavelength. Therefore, a contactless QE spectrum can be constructed from the remotely induced EL superposition of signals. In addition, a variable DC bias can be added so that QE can be measured over a range of forward bias. The DC bias is induced in a third area of the specific device. One advantage of this technique is that only excess carriers that are active in producing the EL signal. These carriers have been labeled as "voltage dependent carriers". This method excludes those carriers that have been generated but diffuse out of the active junction area. These carriers may be detected by PL or PC but do not contribute to the photovoltaic activity of the device.

10.9. Wavelength Resolved PL Spectroscopy

A very mature technology has been employed for many decades of research to analyze the electronic structure of a new material. The bandgap of a newly synthesized semiconductor is typically found by using optical excitation (typically a laser) and finding the most prominent optical emission that is emitted from the material. The intensity of the principal emission is being used as an indicator of material quality. This is perhaps an anecdotal metric, but nevertheless useful. Another very mature diagnostic tool is to optically generate electron-hole pairs at cryogenic temperatures. The excitons produced will bond to impurities producing a unique sub-bandgap emission band that is specific to the impurity. Examples of both applications will be shown in the work that follows.

Thin-film, lattice-matched (LM) $InAs_yP_{1-y}/In_xGa_{1-x}As$ double heterostructures (DHs) were grown using lattice mismatched (LMM) on InP substrates using atmospheric-pressure metalorganic vapor-phase epitaxy (APMOVPE). The bandgap of $In_xGa_{1-x}As$ layers were measured by PL spectroscopy.[15] The room- temperature bandgaps ranged from 0.47 to 0.6 eV.

The bandgaps of the devices were measured by Fourier transform photoluminescence (FTPL) in Fig. 10.17. Curve A is the emission spectrum of a LM DH grown directly on an InP substrate. These epitaxial films were grown at NREL by MOCVD.[16] We find a Gaussian-like spectrum and we can find the bandgap by a linear fit to the square of the near-band-edge portion of the photoluminescence intensity data. Curve B represents the emission spectra of a LMM DH with $E_g = 0.60$ eV. Finally, Curve C represents the emission spectra of a LMM DH with $E_g = 0.48$ eV. The emission spectra of the LMM DHs are relatively strong, indicating that radiative recombination is not quenched by SRH defects.

10.10. Impurity Spectroscopy

Low temperature photoluminescence spectra also produce information about chemical impurities and mechanical defects in the sample. These often produce sub-bandgap emission peaks. In addition,

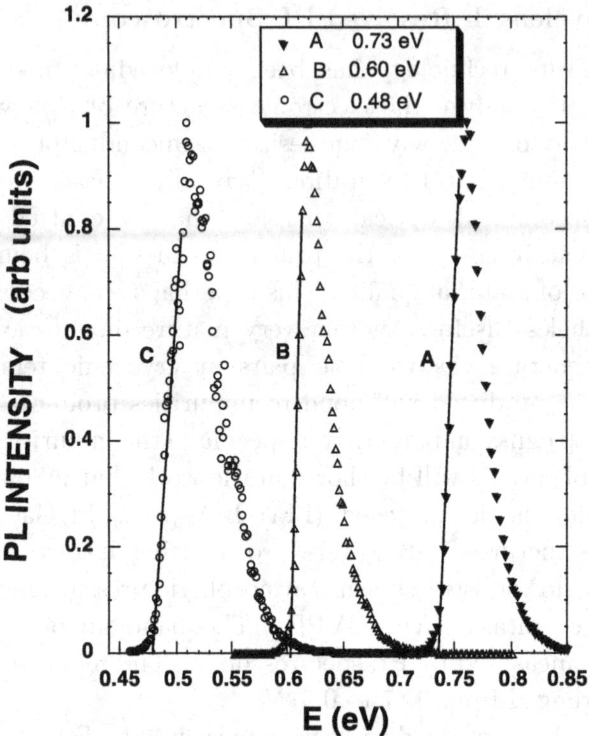

Fig. 10.17. The band-to-band emission peaks of three alloys in the $In_xGa_{1-x}As$ series. The PL peaks are used as an indicator of bandgap. [Reprinted from Ahrenkiel, R. K., Johnston, S. W., Webb, J. D., Gedvilas, L. M., Carapella, J. J., and Wanlass, M. W. (2001). *Appl. Phys. Lett.*, **78**, 1092, with permission of AIP Publishing.]

mechanical quality and strain impact the width of intrinsic band-to-band or exciton peaks.

The data of Fig. 10.18 were obtained[17] at 4.2 K using the excitation from an Ar laser producing 150 mW(cw) and operating at 514.5 nm. A double grating spectrometer with 50 mm slits provided very high wavelength resolution. The subband gap PL peaks have been identified in the EMIS data review. These are primarily excitons bound to phosphorous donors, boron acceptors, or complexes of the latter. These are identified as phonon assisted transitions as identified in the figure. By using a calibrated reference sample, the

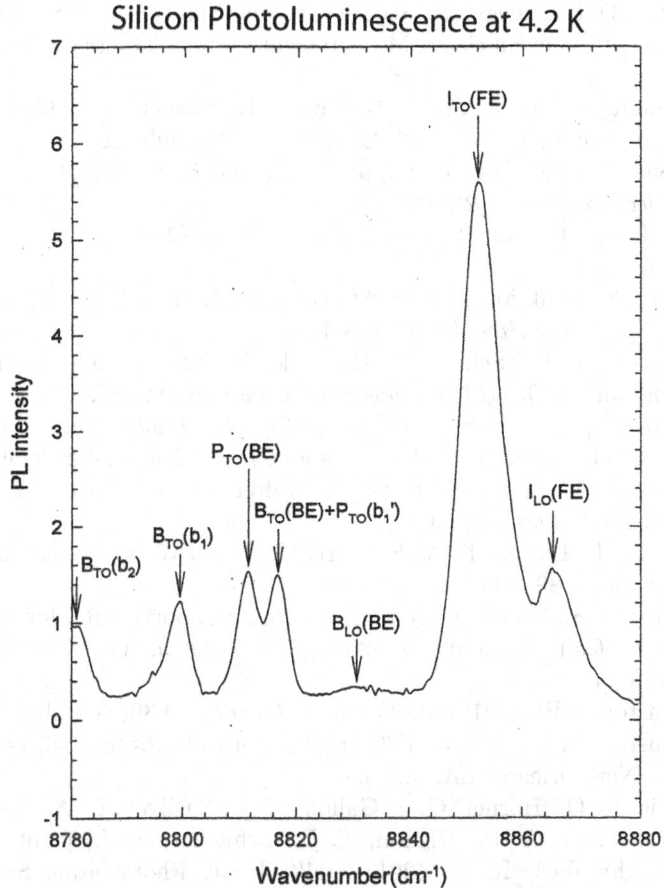

Fig. 10.18. The 4.2 K photoluminescence spectra of a very high quality, float-zone grown silicon wafer. The origins of the various exciton peaks are well-documented in the literature. [Reprinted with permission of The Electrochemical Society. Wang, T., Ciszek, T., and Ahrenkiel, R. K. Electrochemical Society Proceedings V. 96, p. 13. (Electrochemical Society Proceedings Volume 96-12, p. 462).]

low temperature PL intensity has been used to quantify very low concentrations of phosphorous and boron.

References

1. Green, M. A. (1995). *Silicon Solar Cells*, University of New South Wales.

2. Young, D. L., Pinegar, S., Stradins, P., and Egaas, B. (2008). New Real-Time Quantum Efficiency Measurement System, 33rd IEEE PVSC, San Diego, CA.
3. Schmidt, J., Vasilyev, L. A., Hudson, J. E., Horner, G. S., Good, E. A., and Dybiec, M. (2010). 35th IEEE PVSC, Honolulu, HI, pp. 1710–1714.
4. M'Saad, H., Michel, J., Lappe, J. J., and Kimerling, L. C. (1994). *J. Electron. Mater.* **23**, 487.
5. Trupke, T., Bardos, R. A. and Abbott, M. D. (2005). *Appl. Phys. Lett.* **87**, 184102.
6. Wang, X., Juhl, M., Abbott, M., Hameiri, Z., Yao, Y., and Lennon, A. (2014). *Energy Procedia* **55** 169–178.
7. Zwicker, H. R., Scifres, D. R., Holonyak, N., Dupuis, R. D., and Burnham, R. D. (1971). *Solid State Commun.*, **9**, 587.
8. Ahrenkiel, R. K. (1988). In *Current Topics in Photovoltaics, Volume 3*, pp. 44–50. Edited by T. J. Coutts and J. D. Meakin, Academic Press.
9. Giesecke, J. A., Schubert, M. C., Walter, D. and Warta, W. (2010). *Appl. Phys. Lett.* **97**, 092109.
10. Trupke, T., Bardos, R. A., Shubert, M. C. and Warta, W. (2006). *Appl. Phys. Lett.*, **89**, 044107.
11. Johnston, S., Unold, T., Repins, I., Sundaramoorthy, R., Jones, K. M., To, B., Call, N., and Ahrenkiel, R. (2010). *J. Vac. Sci. Technol.*, **A28**, 665.
12. Johnston, S.W. (2012). *J. Vac. Sci. Technol.*, **A30**(4), Jul/Aug.
13. Johnston, S. W. (2015). 42nd IEEE Photovoltaics Specialists Conference, New Orleans, LA, p. 1–6.
14. Davis, K. O., Horner, G. S. Gallon, J. B., Vasilyev, L. A., Lu, K. B., Dirriwachter, A. B., Rigdon, T. B., Schneller, E. J., Öğutman, K., and Ahrenkiel, R. K. (2017). 44th IEEE Photovoltaic Specialists Conference, Washington, DC.
15. Ahrenkiel, R. K., Ellingson, R., Johnston, S., and Wanlass, M. (1998). *Appl. Phys. Lett.*, **72**, 3470.
16. Ahrenkiel, R. K., Johnston, S. W., Webb, J. D., Gedvilas, L. M., Carapella, J. J., and Wanlass, M. W. (2001). *Appl. Phys. Lett.*, **78**, 1092.
17. Wang, T., Ciszek, T., and Ahrenkiel, R. K. Electrochemical Society Proceedings V. 96, p. 13. (Electrochemical Society Proceedings Volume 96-12, p. 462).

CHAPTER 11

Free Carrier Absorption

Another technique for measuring the carrier lifetime is to monitor the transient decay of excess carrier density using a pump-probe method. Free carrier absorption is related to intra-band absorption by the free carriers that are produced by doping or by injection. The free carrier absorption produces a broad absorption spectrum that extends from the band-to-band absorption edge into the infrared spectral region. When excess carriers are produced by an optical pump (or other methods), the free carrier absorption rises and then decreases with the carrier lifetime. The transient free carrier transmission is usually monitored with a steady state infrared light source. Here, we derive the free carrier absorption coefficient using the Drude model of carrier scattering. This will be followed by some measurement data on a number of materials.

There are extensive discussions of the free carrier effects in the literature. One of the classic papers by D. K. Schroder[1] focuses on the free carrier effect in silicon.

11.1. Semi-classical Derivation of the Free Carrier Absorption Coefficient

When free electrons or holes are present, we can use the semiclassical approximation to describe the interaction of the charged carriers with the electric field in the EM wave. We used the Drude model but replaced the discrete scattering by a constant frictional force to produce a uniform velocity in a static electric field. We will write the equation of motion of a free electron in an ac field by introducing a frictional force to provide the effect of scattering:

$$m^* \frac{d^2x}{dt^2} = qE_0 e^{i\omega t} - \gamma \frac{dx}{dt}; \tag{11.1}$$

When the frequency is zero, one gets the Drude equation if:

$$V_d = \frac{qE_0}{\gamma} = \frac{q\tau E_0}{m^*}; \qquad \gamma = \frac{m^*}{\tau}. \tag{11.2}$$

Here, τ is the scattering time from Drude theory.

$$m^* \frac{d^2x}{dt^2} = qE_0 e^{i\omega t} - \frac{m^*}{\tau}\frac{dx}{dt}; \tag{11.3}$$

In equilibrium, the electron (or holes) are positioned so as to produce overall charge neutrality. An electric field induced polarization by displacement of the mobile carriers relative to the fixed donors and acceptors. We will describe this displacement as:

$$x = x_0 e^{i\omega t}; \qquad x_0 = \frac{qE_0}{-m^*\omega^2 + i\frac{m^*\omega}{\tau}}. \tag{11.4}$$

The induced polarization is then written:

$$P = qNx(t) = \frac{-q^2 N E_0 e^{i\omega t}}{m^*\omega\left(\omega - \frac{i}{\tau}\right)}. \tag{11.5}$$

Here, N is the free carrier concentration. From basic electrostatics:

$$\tilde{\varepsilon}E = \varepsilon_0 E + \tilde{P}; \qquad \frac{\tilde{\varepsilon}}{\varepsilon_0} = 1 + \frac{\tilde{P}}{\varepsilon_0 E}. \tag{11.6}$$

The absorption coefficient is contained in the imaginary component of the complex dielectric constant.

$$\frac{\widetilde{\varepsilon}}{\varepsilon_0} = 1 - \frac{q^2 N}{m^* \varepsilon_0 \omega \left(\omega - \frac{i}{\tau}\right)}. \tag{11.7}$$

We will define a scattering frequency as the inverse of the scattering time:

$$\omega_s \equiv \frac{1}{\tau_s}. \tag{11.8}$$

The plasma frequency is defined as:

$$\omega_p \equiv q\sqrt{\frac{N}{\varepsilon_0 m^*}}. \tag{11.9}$$

The complex dielectic constant can be rewritten in terms of these two frequencies:

$$\frac{\widetilde{\varepsilon}}{\varepsilon_0} = 1 - \frac{\omega_p^2}{\omega^2 - i\omega_s\omega}. \tag{11.10}$$

We can rationalize the complex expression to produce a real and imaginery component.

$$\frac{\widetilde{\varepsilon}}{\varepsilon_0} = 1 - \frac{\omega_p^2}{\omega^2 + \omega_s^2} + i\frac{\omega_s}{\omega}\frac{\omega_p^2}{\omega^2 + \omega_s^2}. \tag{11.11}$$

The dielectric coefficient is expression as real and imaginery components ε_1+iε_2.

$$\begin{aligned} \varepsilon_1 &= n_{op}^2 - k^2 = \varepsilon_0 \left(1 - \frac{\omega_p^2}{\omega^2 + \omega_s^2}\right); \\ \varepsilon_2 &= 2n_{op}k = \varepsilon_0 \left(\frac{\omega_s}{\omega}\frac{\omega_p^2}{\omega^2 + \omega_s^2}\right). \end{aligned} \tag{11.12}$$

Here, n_{op} is the optical index or refraction and λ is the wavelength of the incident optical beam.

The Beers law free carrier absorption coefficient is then derived from the above as;

$$\alpha = \frac{4\pi_{op}k}{\lambda} = \frac{2\pi}{n_{op}\lambda}\varepsilon_0\left(\frac{\omega_s}{\omega}\frac{\omega_p^2}{\omega^2+\omega_s^2}\right). \tag{11.13}$$

$$\alpha = \frac{q^3N}{2\pi(m^*)^2 n_{op}c\mu}\frac{1}{\omega^2+\omega_s^2}. \tag{11.14}$$

This can be expressed in terms of more common parameters:

$$\alpha = \frac{q^3N}{2\pi(m^*)^2 n_{op}c\mu}\frac{1}{\omega^2+\omega_s^2}. \tag{11.15}$$

The optical frequencies for semiconductor absorptions are in the range of $10^{15}\,\mathrm{s}^{-1}$ whereas the scattering frequency is in the range $10^{12}\,\mathrm{s}^{-1}$.

$$\omega^2 \gg \omega_s^2.$$

Therefore, one can calculate the free carrier absorption as:

$$\alpha = \frac{q^3N\lambda^2}{4\pi^2\varepsilon_0(m^*)^2 n_{op}c\mu} = C_1\frac{N\lambda^2}{\mu} \tag{11.16}$$

Here, N is carrier concentration, λ is the incident wavelength, and μ is the carrier mobility. C_1 contains fundamental constants and the carrier effective mass.

At a given probe wavelength, the free carrier absorption varies linearly with the scattering frequency (or inversely with mobility). The fundamental absorption process involves the transfer of optical energy to phonons of the host material via the scattering of free carriers. In the Drude theory, we assumed that the scattering lifetime, τ, was a constant. In real materials, the scattering time is a function of carrier concentration and temperature. Therefore, the free carrier absorption varies as:

$$\alpha = C_1\frac{N\lambda^{\mathrm{P}}}{\mu}. \tag{11.17}$$

where p varies 1.5 to 3.5 depending on the dominant phonon scattering mechanism. As the scattering process approaches zero (or mobility becomes infinite), the free carrier absorption approaches zero and the material is transparent.

The interesting contrast with photoconductive decay is that the signal is inversely proportional to mobility as compared to linearly increasing with mobility in PCD. This difference will be utilized in a later section dealing with dual techniques.

11.2. Experimental Data

There are abundant examples of data in the literature that exhibit the free carrier absorption effect. Schroder fits existing data and quotes the free carrier absorption coefficients in silicon as:

$$\begin{aligned} \alpha_n &\approx 1 \times 10^{-19}\lambda^2 n; \\ \alpha_p &\approx 2.7 \times 10^{-18}\lambda^2 p. \end{aligned} \tag{11.18}$$

Here n is the electron concentration and p is the hole concentration, and λ is the wavelength in microns. The absorption coefficient for holes is larger than for electrons reflection the inverse mobility ratio of the two charge carriers. The absorption coefficients for FC studies are quite weak compared to the band-to-band coefficients. For example, with an electron concentration of $1 \times 10^{17}\,\text{cm}^{-3}$, the FC absorption coefficient at $10.6\,\mu\text{m}$ (the wavelength of a CO_2 laser), the computed value of α is: $1.1\,\text{cm}^{-1}$. Therefore, the FC absorption is primarily useful for thick wafers and not for thin films.

One of the early and classic examples is shown for the free carrier absorption in GaAs[2] as a function of doping level. The doping level ranges from undoped ($5 \times 10^{14}\,\text{cm}^{-3}$, Sample #1) to $5.4 \times 10^{18}\,\text{cm}^{-3}$ (sample #6). The doping concentrations of the six samples are shown in Table 11.1.

The absorption coefficient increases as λ^2 with doping level as predicted by the theory. The spectral region between the absorption edge and about four microns is dominated by transitions that are unrelated to free carrier absorption.

Analysis shows that there is a minimum in the reflection coefficient at the plasma resonance. Using the reflectance data and knowing the carrier concentration, the carrier effective mass can be

Table 11.1. The concentration for the samples providing the data of Fig. 11.1.

Sample	Concentration (cm^{-3})
1.	$<5 \times 10^{15}$
2.	1.3×10^{17}
3.	4.9×10^{17}
4.	1.09×10^{18}
5.	1.12×10^{18}
6.	5.4×10^{18}

computed. This has been reported for a number of materials by the early work of Spitzer[4] and coworkers as shown in Fig. 11.1. The absorption coefficient varies as wavelength, $\lambda 1.5$. The free carrier absorption spectrum of germanium at 4.2 K is shown in Fig. 11.2. The data of Fig. 11.3 shows infrared reflectance from doped germanium. The reflectance minimum occurs at the plasma frequency that changes with doping level.

11.3. Transient Free Carrier Decay

When excess carriers are produced by absorbed photons, the infrared transmission will be reduced by free carrier absorption. The carrier lifetime has been measured by a pump-probe technique using a cw infrared source to monitor the transient transmission as the excess carriers decay in real time.

A schematic diagram of a transient free carrier decay (FCD) apparatus is shown in Fig. 11.4. This system uses a cw carbon dioxide laser as a probe, and the relative transmission is plotted versus time. The series of emission bands that are centered at about 10.6 μm were used in a series of experiments to be described later in this section. We can write the absorption coefficient of a silicon wafer at 10.6 μm using the absorption coefficients of Schroder and coworkers as:

$$\begin{aligned} \alpha_n &\approx 1.2 \times 10^{-17}(N + \Delta n) \equiv c_n n; \\ \alpha_p &\approx 3.0 \times 10^{-16}(P + \Delta p) \equiv c_p p. \end{aligned} \tag{11.19}$$

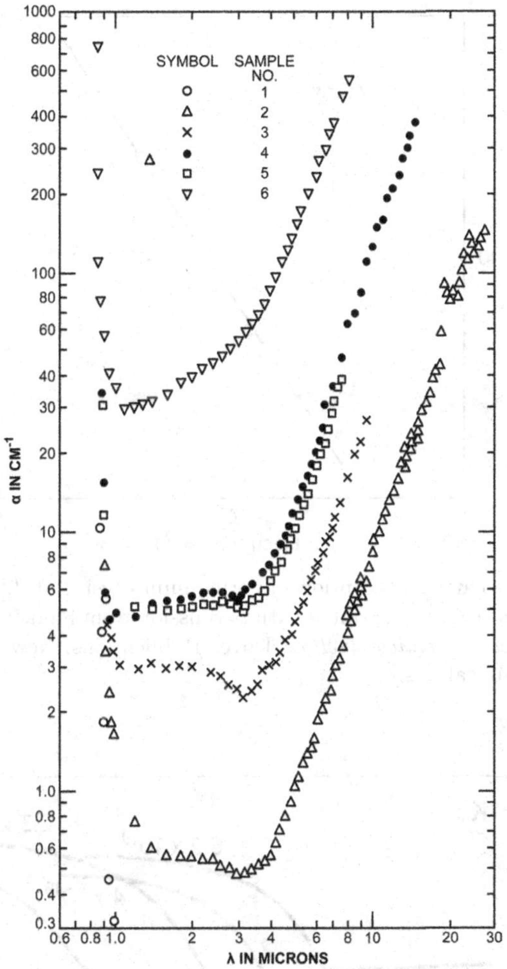

Fig. 11.1. Free carrier absorption spectra of doped GaAs from Spitzer and Whelan.[1] [© [2019] IEEE. Reprinted, with permission, from Schroder, D. K. (1978) *IEEE Trans. on Electron Devices* ED-25.]

The steady state probe transmission of a p-type silicon wafer that has thickness d and a doping concentration N in the pump-probe system below is;

$$I_{ss} = I_0 (1 - R)^2 \exp(-c_p N d); \qquad (11.20)$$

Here R is the reflection loss at the entrance and exit surfaces. When Δn excess electrons are injected into the wafer, the FC absorption

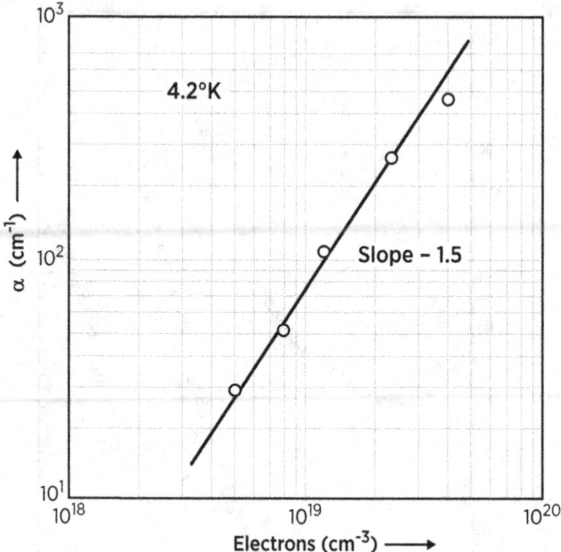

Fig. 11.2. Free carrier absorption of germanium crystal at 4.2 μm wavelength and 4.2 K temperature.[3] [Reprinted with permission from Pankove, J. I. (1971). In *Optical Processes in Semiconductors*, Dover Publications, New York. Copyright [2019] Dover Publications.]

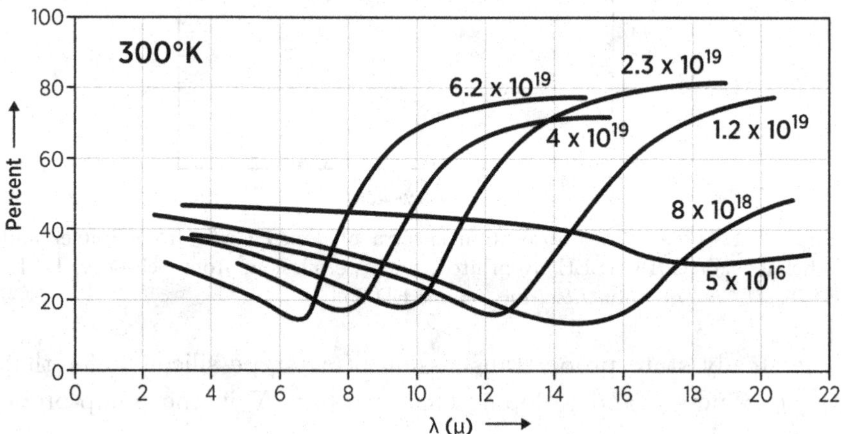

Fig. 11.3. The reflectance spectra[3] of germanium with six different doping levels. The reflectance minima corresponds to the plasma frequency at each doping level. [Reprinted with permission from Spitzer, W. G. and Fan, H. Y. *Phys. Rev.* **106**, 882 (1957). Copyright (2019) by the American Physical Society.]

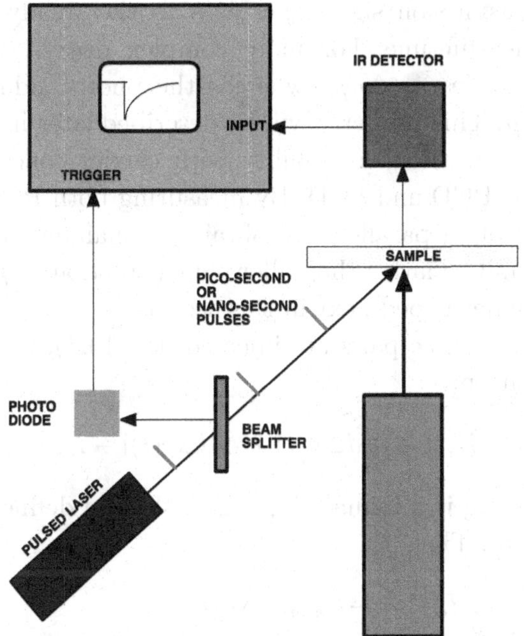

Fig. 11.4. Schematic of a typical apparatus for measuring the transient free carrier absorption using a pulsed short wavelength laser as a pump and a CO_2 cw laser as a probe.

increase, with a decrease in the cw transmission signal.

$$\begin{aligned} I_t(t) &= I_0(1-R)^2 \exp(-c_p N - c_n \Delta n(t))d; \\ &= I_{ss} \exp(-c_n \Delta n(t) d). \end{aligned} \tag{11.21}$$

Typical values for a silicon wafer are $d = 300\,\mu\text{m}$, and $\Delta n = 1 \times 10^{16}\,\text{cm}^{-3}$. In this typical case, the argument $c_n^* \Delta n^* d$ is calculated to be 0.09. Therefore, the exponential may be expanded:

$$\begin{aligned} \Delta I_t(t) &= I_t(t) - I_{ss} = I_{ss}(1 - c_n d \Delta n(t)) - I_{ss}; \\ &= -I_{ss} c_n d \Delta n(t). \end{aligned} \tag{11.22}$$

Assuming that the excess electrons decay exponential in time, the transmitted intensity is:

$$\Delta I_t(t) = -I_{ss} c_n d \Delta n_0 \exp(-t/\tau), \tag{11.23}$$

The probe transmission signal increases to the steady state value with the carrier lifetime. For more complex decays, we see that the FC signal varies as $\Delta n/\mu$ whereas the photoconductive signal decays as $\Delta n^*\mu$. This property will be described later in this section to analyze the variation of mobility with carrier concentration by doing combined PCD and FCD. By measuring both PCD and FCD simultaneously or in parallel, one obtains a unambiguous values of $\Delta n(t)$ and $\mu(t)$. The latter then allows a calculation of $\mu(\Delta n)$. This method will be developed next in this section.

When electron-hole pairs are injected, the 10.6 μm transmission transient is written:

$$\Delta I_t(t) = I_{ss}[1 - c_n d\Delta n(t) - c_p d\Delta p(t)] - I_{ss}. \tag{11.24}$$

If shallow trapping is minimal, $\Delta n = \Delta p$. We can define c_{np} as; $c_{np} \equiv c_n + c_p$. Then:

$$\Delta I_t(t) = -I_{ss}c_{np}d\Delta n(t). \tag{11.25}$$

$$\Delta I_t(t) = -I_{ss}c_{np}d\Delta n_0 \exp(-t/\tau), \tag{11.26}$$

For the *e-h* pairs recombination, there is a larger drop in the 10.6 μm probe transmission and the transmission tracks the recombination lifetime.

Figure 11.5 shows free-carrier transient decay in a *p*-type silicon wafer using a 1064 YAG laser pump and cw-CO_2 probe measured by Feldman[5] and coworkers. The float-zone grown wafer was unpassivated and had a background doping level of $1.5 \times 10^{14}\,\text{cm}^{-3}$. The pump injection levels are indicated in the figure. These data were measured using a dual sensor apparatus that is described in the next section. The data show a well-behaved exponential decay under these low injection conditions.

The FCA technique is really only applicable to thick wafers because the free carrier absorption is very small and requires thicker samples to acquire credible data.

Linnros[6] reported one of the first extensive studies on free carrier decay analysis in 1998. He used a 1.06 μm YAG pulse and a 3.39 μm HeNe probe to study free carrier transients in silicon. The

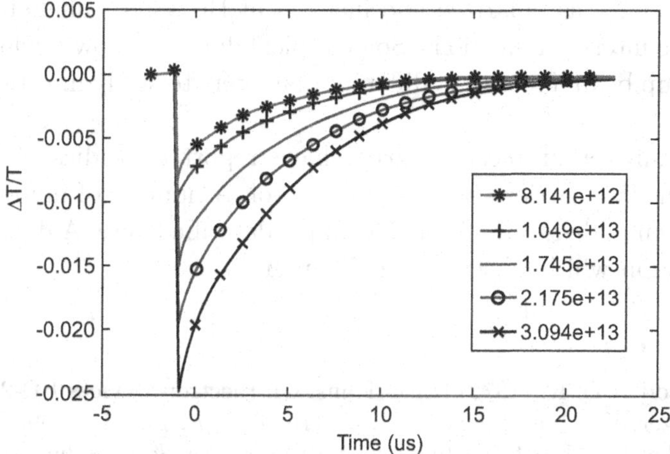

Fig. 11.5. Transmission free carrier decay data using a 1064 nm pulsed YAG pump and a cw-CO_2 probe measured by Feldman and coworkers on a silicon wafer. The pump injection levels used for each decay curves are shown. [Reprinted from Linnros, J. (1998) *J. Appl. Phys.* **84**, 275, with permission of AIP Publishing.]

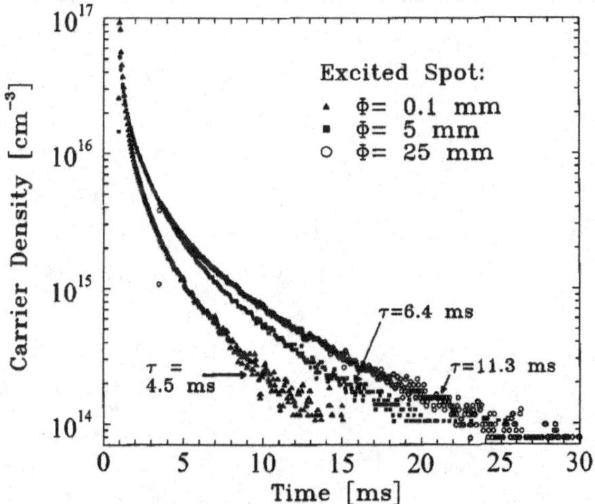

Fig. 11.6. Data from the work of Linnros with pump spot size as a parameter. Here, the excess carrier density is plotted as a function of time. The variations in decay time are attributed to lateral diffusion out of the initial excitation volume. The transmitted signal is converted to free carrier density as a function of time. [Reprinted from Linnros, J. (1998) *J. Appl. Phys.* **84**, 275, with permission of AIP Publishing.]

probe beam was polarized and incident at Brewster's angle to avoid multiple internal reflections. Some typical data are shown below with the pump beam focused to different diameters to study injection level effects.

A number of recent journal have reported studies using the transient free carrier absorption as a tool. A number of innovations[7] have been added to the basic technique described here. A dual sensor application will be discussed in Chapter 14.

References

1. Schroder, D. K. (1978) IEEE Trans. On Electron Devices ED-25.
2. Spitzer, W. G. and Whelan, J. M. (1959) *Phys. Rev.* **114**, 59.
3. Pankove, J. I. (1971). In *Optical Processes in Semiconductors*, Dover Publications, New York.
4. Spitzer, W. G. and Fan, H. Y. (1957) *Phys. Rev.* 106, 882.
5. Feldman, A., Ahrenkiel, R., and Lehman, J. (2013) *J. Appl. Phys.* **113**, 103703.
6. Linnros, J. (1998) *J. Appl. Phys.* **84**, 275.
7. Zhang, X., Li, B., and Gao, Gao (2008) *J. Appl. Phys.* **103**, 033709.

CHAPTER 12

Recombination in Charge Separation Structures

Charge separation occurs in structures that develop an internal electric field due to doping or composition discontinuities. The electric field separates the electron-hole pairs and produces charge separation. This usually means that the two carrier types become majority carriers as they transfer to *n*- and *p*-type regions because of the internal field. Typical devices containing such *E*-fields are *pn* junctions and heterostructures. In the case of the type II heterostructure, the internal field at the interface separates charge because of the unique band alignment.

The confinement structure does not allow carriers to flow out of the volume under test, and therefore the carrier decay process is pure recombination. The measurement of excess carrier lifetime may be measured by transient luminescence (TRPL) in direct bandgap semiconductors. The decay time tracks the excess carrier density as a function of time and allows a direct measurement of the carrier lifetime. This method uses algorithms to extract a lifetime from the intensity of the steady state intensity. When carriers flow out of the volume of interest, more elaborate methods are required to extract a lifetime from the data. The interpretation of the data is more complicated for *pn* junction structures that will be the topic of this section. The homojunction or heterojunction is a basic structure in devices ranging from diode lasers to photovoltaic cells. Direct measurement of the minority-carrier lifetime of a junction device may be the only method of analysis available. Here, models of junction excess carrier decay have been made to account for effects of the junction on measurements.

In the *pn* junction, the internal electric field drives holes to the *p*-region and electrons to the *n*-region. Consequently, the PL and PC signals will differ significantly. The PL signal varies as;

$$\begin{aligned} I_{\mathrm{PL}} &= BN_{\mathrm{A}}\Delta n(t); \quad p - regions; \\ or&; \\ I_{\mathrm{PL}} &= BN_{\mathrm{D}}\Delta p(t). \quad n - regions. \end{aligned} \tag{12.1}$$

The minority carrier in either region will diffuse to the space charge region and drift rapidly to the majority region. The diffusion process terminates the PL signal but the photoconductivity continues until the separated carriers either flow through external circuitry or decay by forward injection into the junction.

$$\Delta\sigma(t) = q\mu_n\Delta n(t) + q\mu_p\Delta p(t); \tag{12.2}$$

If there is an external electrical path between the emitter and base, a current will flow in the external circuit. In that case, this device is a photo diode. If the external contacts are electrically isolated, an open-circuit voltage will develop. The device now operates in the photovoltaic mode and a current will flow in the forward direction across the depletion region as the majority charge increases. The V_{oc} increases and becomes a forward biased diode. In either case, the time dependence of the TRPL signal will differ significantly from the PCD signal.

The first author analyzed the PL decay of thick emitter, *pn* thin-film devices, using the configuration in Fig. 12.1 and using two different analytical approaches.[1] The approaches were the Laplace transform method and the modal series methods developed in Chapter 5. For these calculations, the simplified band diagram of a

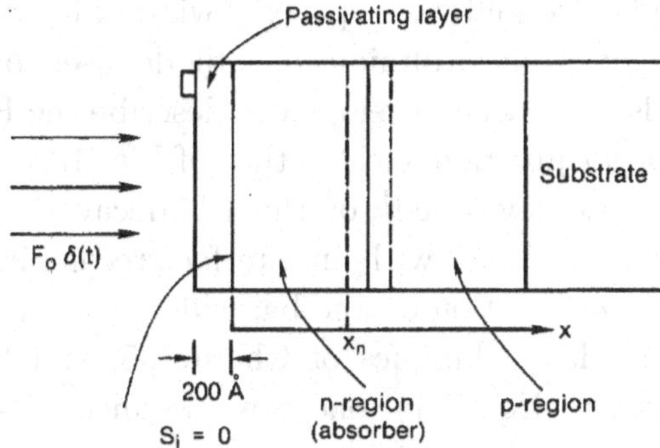

Fig. 12.1. Model of the device used in the modal calculations of PL decay.

PN junction was the basis for the calculation of the TRPL. After an instantaneous generation of electron-hole pairs, the photoelectrons diffuse to the junction (depletion region) where they drift in the high electric field to the n-type base. It was shown that the long-term, transient PL decay time was derived by the modal technique (Chapter 5) and found to be approximately:

$$\frac{1}{\tau_{pL}} = \frac{1}{\tau} + \frac{\pi^2 D}{4X_{\mathrm{J}}^2};$$

where D is the electron diffusion coefficient and τ is the bulk lifetime. The total decay time becomes smaller than the recombination time and decreases as the inverse square of the length of the film. One can say that charge separation "quenches" the PL decay time. If the junction is shorted or has reverse bias, the PL signal terminates when the minority carrier diffusion is complete. When the injection level is very small, the boundary condition at X_{J} is approximately correct, even if there is no external shorting. In that case, the photoconductivity increases until the excess majority charge is injected back into the junction after optical generation terminates. One observes majority carrier photocurrent until the V_{oc} discharges internally.

Figure 12.1 shows a schematic diagram of a device structure that was used for this calculation. The figure shows an emitter and a base of an np junction device. For GaAs devices, the emitter is usually passivated by a lattice-matched, epitaxial window layer. The junction is a component of a photovoltaic cell, a diode laser, or a transistor. Several methods have been developed to describe the PL decay from the emitter of a *pn* junction such as that of Fig. 12.1.

A solution was developed[1] of the PL decay in this structure bounded by a window layer with an interface recombination velocity $S_I = 0$ and by a *pn* junction on the backside.

Using the modal techniques of Chapter 5, and the eigenvalue equation, derived in Eq. (5.7), one can use unequal recombination velocities at the two interfaces:

$$\tan \alpha_n X_J = \frac{\alpha_n X_J (S_w + S_J) X_J / D}{(\alpha_n X_J)^2 - S_w S_J (X_J/D)^2} \tag{12.3}$$

With S_W finite and with S_J approaching infinity, the equation becomes:

$$\tan \alpha_n X_J = -\frac{\alpha_n X_J}{S_w X_J / D} \tag{12.4}$$

Letting $\alpha_n X_J$ equal θ_n and $S_w\ X_J/D$ equal z, this equation becomes:

$$\tan \theta_n = -\frac{\theta_n}{z} \tag{12.5}$$

One can evaluate these solutions for specific device parameters. The time decay is written as a sum of modes with a specific lifetime for each mode. The mode lifetime was shown in Eq. (5.17):

$$\frac{1}{\tau_n} = \frac{1}{\tau} + \beta_n = \frac{1}{\tau} + D\alpha_n^2. \tag{12.6}$$

An analytic solution to this problem exists for the case $S \approx 0$ and $S_J \to \infty$. In this case, the eigenvalues of θ_n are:

$$\begin{aligned} &\theta_n = (2n-1)\frac{\pi}{2}; \\ &n = 1, 2, 3, \ldots. \\ &\theta_n = \pi/2, 3\pi/2, 5\pi/2, \ldots \quad \text{for } n = 1, 2, 3, \end{aligned} \tag{12.7}$$

The corresponding values of β_n are for $n = 1, 2, 3$, etc.:

$$\beta_n = D\frac{(2n-1)^2\pi^2}{4X_\mathrm{J}^2} \tag{12.8}$$

The one-dimensional function $U(x,t)$, is

$$U(x,t) = \sum_{n=1}^{\infty} A_n \cos\left[(n-1/2)\frac{\pi X}{X_\mathrm{J}}\right] e^{-\beta_n t} \tag{12.9}$$

For an initial Beers' law distribution, the coefficients A_n are found by integrating:

$$\int_0^{X_\mathrm{J}} \alpha I_0 \exp(-\alpha x) U(x,0) dx;$$

$$A_n = I_0 \frac{2\alpha^2 X_\mathrm{J} + 2\pi^2\alpha(-1)^{n+1}(n-1/2)\exp^{-\alpha X_\mathrm{J}}}{\alpha^2 X_\mathrm{J}^2 + \pi^2(n-1/2)^2}. \tag{12.10}$$

In the limit of $\alpha x_\mathrm{J} \gg 1$, one finds $A_n = 2I_0/x_\mathrm{J}$. Using this result, one finds the total excess charge density by integrating $\rho(x,t)$ from 0 to x_J:

$$\begin{aligned} \rho_t(t) &= \exp(-t/\tau)\int_0^{x_n} U(x,t)dx \\ &= \frac{4I_0}{\pi}\sum_{n=1}^{\infty}\frac{(-1)^{n-1}}{2n-1}\exp\left[-\left(\beta_n + \frac{1}{\tau}\right)t\right]. \end{aligned} \tag{12.11}$$

Thus:

$$I_\mathrm{PL}(t) = \frac{4I_0}{\pi\tau_\mathrm{R}}\sum_{n=1}^{\infty}\frac{(-1)^{n+1}}{2n-1}\exp\left[-\left(\beta_n + \frac{1}{\tau}\right)t\right]. \tag{12.12}$$

This series converges rapidly for $t \gg \frac{x_\mathrm{J}^2}{\pi^2 D}$ as the first mode ($n = 1$ term) dominates:

$$I_\mathrm{PL}(t) = \frac{4I_0}{\pi\tau_R}\exp\left[-\left(\frac{\pi^2 D}{4x_\mathrm{J}^2} + \frac{1}{\tau}\right)t\right]. \tag{12.13}$$

The term $\frac{x_J^2}{D\pi^2}$ was defined previously as the diffusion transit time τ_D. Therefore:

$$\frac{1}{\tau_{PL}} = \frac{1}{\tau} + \frac{1}{4\tau_D} = \frac{1}{\tau} + \frac{\pi^2 D}{4x_J^2}. \quad (12.14)$$

12.1. Time Resolved Photoluminescence of PN Junctions

In the n-type emitter, the PL intensity is given by:

$$I_{PL}(t) = B(\Delta n(t)^* N_D + \Delta n(t)\Delta p(t)). \quad (12.15)$$

The population of excess electrons in the base, during short circuit conditions, decreases as:

$$\Delta n(t) = \exp(-t/\tau)^* \exp(-t/\tau_D). \quad (12.16)$$

When the recombination lifetime is much longer than the diffusion transit time, $t \gg \tau_D$, the observed TRPL signal is:

$$I_{PL}(t) \cong BN_A \exp\left(-\frac{4d^2}{\pi^2 D}t\right). \quad (12.17)$$

In this case, the TRPL signal is dominated by diffusion transit rather than recombination. This device may be designed to measure the minority carrier mobility rather than lifetime by making diffusion transit the dominant decay process.

The calculated transient PL decay was computed and shown in Fig. 12.2 as a function of minority carrier mobility. This applies to the case of a p-type emitter and n-type base. As the mobility increases, the duration of the decay time decreases, and the charge separation occurs more quickly. The PL signals terminates after the charge separation is complete.

Structures have been designed specifically to measure the minority-carrier diffusion time. The emitter or base length is chosen so that the diffusion transit time is much shorter than the recombination lifetime. The charge separation dynamics are measured by either TRPL or by the transient short-circuit current. The method

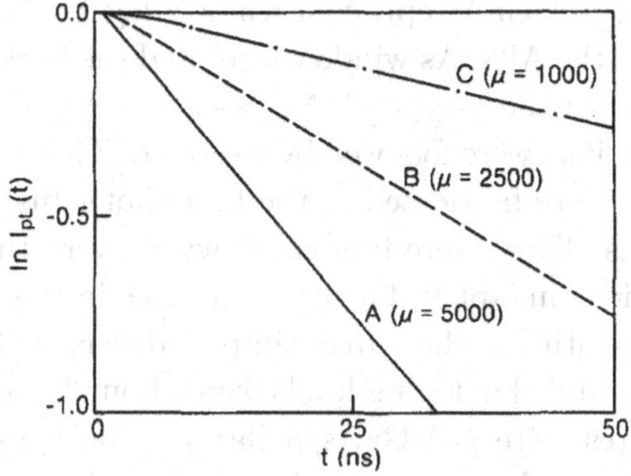

Fig. 12.2. Calculations of the PL decay in the device of Fig. 12.1 with $X_J = 10.0\,\mu$m and $\tau = 1000$ ns. The minority carrier mobility is the variable parameter.

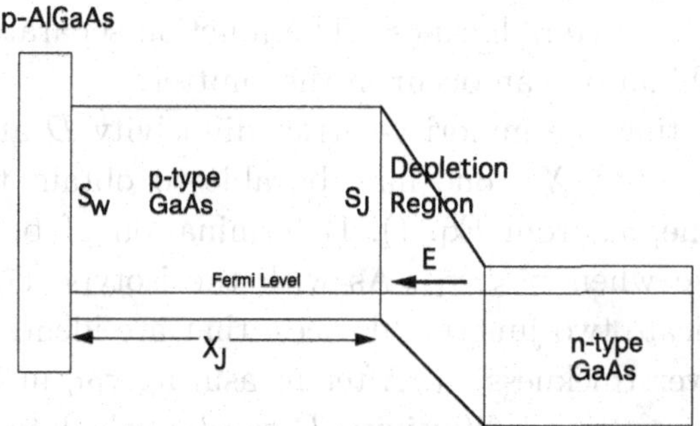

Fig. 12.3. The model device structure used by Ehrhardt and coworkers for the calculations used to model transient PL decay. [Reprinted by permission from Springer Nature, *Appl. Phys.* **A53**, Transient photoluminescence decay study of minority carrier lifetime in GaAs heteroface solar cell structures, Ehrhardt, A., Wettling, W., and Bett, A. (1991) 123–129. Copyright 2019.]

has been labeled as the time-of-flight (TOF) technique by several workers.[2–4] The method is especially useful for thin films as the diffusion coefficient is not easily measured in such devices.

Ehrhardt[5] and coworkers did extensive theoretical and experimental studies in 1991 and illustrated the effects of a back junction of the transient PL decay from the thick emitter. Their device structure that was similar to that of Fig. 12.3 and used their calculations and PL decay measurements. The *p*-GaAs emitter is passivated by a lattice-matched p-AlGaAs epitaxial window layer. The front surface is passivated by the AlGaAs window layer and the back surface forms a pn junction with the n-type GaAs base.

The calculations were followed by a series of TRPL measurements on devices that were fabricated at the Fraunhofer Institute for Solar Energy Systems. These were fabricated with a very thick emitter so that the diffusion time played a significant role in the decay process. The junction depth for the three samples discussed here were 1.9, 2.5, and 3.8 μm and the doping levels were about 1×10^{18} cm^{-3}. The device structures were p-AlGaAs/p-GaAs/n-GaAs/n-AlGaAs. The p-GaAs emitter thickness was varied as described, and the decay time increased with thickness because of the addition of the diffusion transit time. The authors used a Fourier transform method to solve the transport equations. One sees from the figure that the decay time decreases with emitter thickness. The junction separates the charge before recombination can occur in the emitter.

By estimating the minority-carrier diffusivity D and measuring the junction depth X_J, one may be able to obtain the minority-carrier lifetime, τ, from Eq. (). Determination of bulk lifetime is most accurate when $\tau < \tau_D$. As with the isotype DH structures, one can fabricate two junction devices, that are identical except for the active layer thickness x_J. After measuring τ_{PL} in both devices, the bulk lifetime τ and diffusivity D can be calculated. If $\tau \gg \tau_D$, the bulk minority-carrier lifetime is insignificant in decay kinetics and depends only on D:

$$I_{PL}(t) = \frac{4I_0}{\pi\tau_R} \exp\left(-\frac{\pi^2 Dt}{4x_J^2}\right) \tag{12.18}$$

In this case, the photoluminescence lifetime is completely unrelated to the recombination lifetime and depends only on the junction depth and minority-carrier diffusivity. At long times, the first mode dominates the decay process. In Eq. (12.16), β_1 comes from the solution of the eigenvalue equation found in Chapter 5.

The Fraunhofer group applied a Fourier transform method to the one-dimensional, time-dependent diffusion equation. A generation function $\delta(t)\alpha\exp(-\alpha x)$ is used to generate a solution, PL(t), which is mathematically quite complex. The Fourier transform converts the partial differential diffusion equation into an ordinary differential equation of the spatial variable x. Input parameters are the window-emitter recombination velocity S_w the junction recombination velocity S_J, and the emitter junction depth X_J. The value of S_J depends on the bias voltage across the junction. Under reverse bias, S_J is approximated as infinity. This boundary condition requires that $\rho(X_J)$ is zero. Under forward bias, $\rho(X_J)$ is an exponential function of the voltage and the solutions are more mathematically complicated.

Ehrhardt and coworkers used an elegant Fourier transform method to solve the one-dimensional, time-dependent diffusion equation. A generation function $\delta(t)\alpha\exp(-\alpha x)$ was used to generate a solution PL(t) which is mathematically quite complex. The Fourier transform converts the partial differential diffusion equation into an ordinary differential equation of the spatial variable x. Input parameters are the window-emitter recombination velocity S_w, the junction recombination velocity S_J, and the emitter junction depth X_J. The value of S_J depends on the bias voltage across the junction. Under reverse bias, S_J is approximated as infinity. This boundary condition requires that $\rho(X_J)$ is zero. A transcendental equation is derived with a series of terms that correspond to a modal series. The boundary conditions require unequal recombination velocities at the two interfaces:

$$\tan\alpha_n X_J = \frac{\alpha_n X_J(S_w + S_J)X_J/D}{(\alpha_n X_J)^2 - S_w S_J(X_J/D)^2} \tag{12.19}$$

This is the same result that came out of the modal approach resulting in Eq. (5.7). In the limit of S_J approaching infinity, the equation becomes:

$$\tan\alpha_n X_J = -\frac{\alpha_n X_J}{S_w X_J/D} \tag{12.20}$$

Letting $\alpha_n X_J$ equal θ_n and $S_w X_J/D$ equal to z, this equation becomes:

$$\tan\theta_n = -\frac{\theta_n}{z} \tag{12.21}$$

One can evaluate these solutions for specific device parameters. The time decay is written as a sum of modes with a specific lifetime for each mode. The mode lifetime for each mode is equal to that shown in Eq. (5.18):

$$\frac{1}{\tau_n} = \frac{1}{\tau} + \beta_n = \frac{1}{\tau} + D\alpha_n^2 \tag{12.22}$$

The first root lies between $\pi/2 < \theta_n < \pi$; the second root lies between $3\pi/2 < \theta_n < 2\pi$, etc.

Figure 12.4 shows the spatial distribution of $\rho(x,t)$ in a model system at various times after a $\delta(t)$ excitation of minority carriers. The parameters used in the calculation are $X_j = 4.0\,\mu\text{m}$, $\tau = 3\,\text{ns}$,

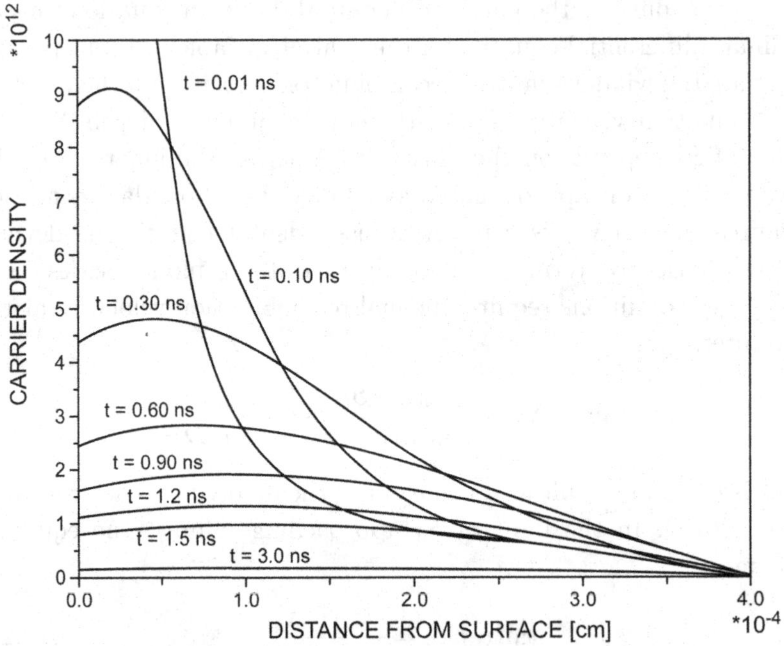

Fig. 12.4. Calculations of Ehrhardt and coworkers for the excess carrier density versus time for the device of Fig. 12.3 with *XJ* equal to $4.0\,\mu\text{m}$. The calculation lifetime value was 3 ns and the diffusivity was $25\,\text{cm}^2/\text{s}$. [Reprinted by permission from Springer Nature, *Appl. Phys.* **A53**, Transient photoluminescence decay study of minority carrier lifetime in GaAs heteroface solar cell structures, Ehrhardt, A., Wettling, W., and Bett, A. (1991) 123–129. Copyright 2019.]

and $S_w = 10^5$ cm/s. As S_w is large, $\rho(0,t)$ bends downward if there is minority-carrier recombination at the interface. At X_J, $\rho(x_J,t)$ is maintained at 0 as S_J is set to the very large value of 10^7 cm/s. The identical solution exists for S_J equal to infinity.

Figure 12.2 shows the calculation of spatial distribution by Ehrhardt *et al.* of $\rho(x,t)$ over a time range from 0 to 3 ns after the $\delta(t)$ excitation pulse. The parameters used in the calculation are $X_j = 4.0\,\mu$m, $\tau = 3$ ns, and $S_w = 10^5$ cm/s. As S_w is large, $\rho(0,t)$ bends downward if there is minority-carrier recombination at the interface. At X_J, $\rho(x_J,t)$ is maintained at 0 as S_J is set to the very large value of 10^7 cm/s. We see from Fig. 12.2 that the excess electrons have almost completely diffused from the emitter in 3 ns. Thus, the PL decay time is much shorter than the 3 ns bulk lifetime because of diffusion and charge separation. The PL decay is "quenched" by the process and the measured lifetime is less than the bulk lifetime. In the Gerhardt experimental work, the back contact was reverse biased to assure that there was no open-circuit voltage buildup during the measurement. Measurements of the transient PCD were not made during these measurements.

If an open circuit conditions exists, the voltage increases to an open-circuit voltage condition. The open circuit rise time is determined by the diffusion transit time of electrons from the base region to the depletion boundary at $x = 0$, or;

Open Circuit V_{oc} of GaAs Thin Film

As a very dramatic example of the differences between PCD and TRPL decay in PN junctions, we show the data of Figs. 12.5, figures A and B. The data were measured on a PN junction device with structure n-AlGaAs/n^+GaAs/p-GaAs/p-AlGaAs. The SRV of the AlGaAs/GaAs interfaces were less than 1×10^3 cm/s. The thickness of the n^+GaAs emitter is 0.1 μm and it doped to a level greater than 1×10^{18} cm^{-3}. The Zn doping of the p-type base is 1.4×10^{17} cm^{-3}. The TRPL lifetime was measured by time-correlated single photon counting and the photoconductive transient was measured by RCPCD. The device is run in an open circuit condition, but the TRPL data was obtained at a very low injection level.

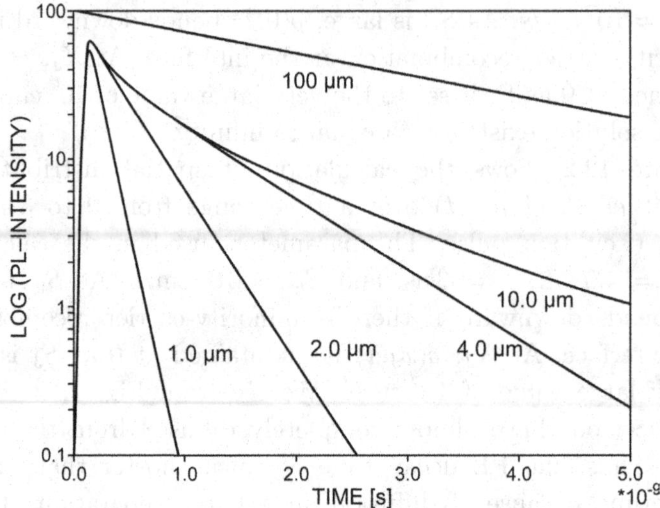

Fig. 12.5. Calculated PL decay[5] using a bulk lifetime of 3 ns and a diffusion coefficient of $25\,\mathrm{cm}^2/\mathrm{s}$. The emitter thickness is varied in the calculation from 1 to 100 μm. [Reprinted by permission from Springer Nature, *Appl. Phys.* **A53**, Transient photoluminescence decay study of minority carrier lifetime in GaAs heteroface solar cell structures, Ehrhardt, A., Wettling, W., and Bett, A. (1991) 123–129. Copyright 2019.]

The radiative lifetime of the base is 28 ns, but the TRPL data of this structure is shown in Fig. 12.7(a) and shows an initial lifetime of 1.4 ns. We will assume that the initial decay is the diffusion transit time in the base as the charges separate after the *ps*-duration light pulse. Using the previous model and assuming short circuit conditions, this value of τ_D (1.4 ns) requires a diffusivity of $11.6\,\mathrm{cm}^2/\mathrm{s}$ and an electron mobility of $463\,\mathrm{cm}^2/\mathrm{Vs}$. Here we assumed that:

$$D_n = \frac{4d^2}{\pi^2 \tau_D}. \tag{12.23}$$

Curve B is a plot of the same PN device using photoconductive decay (RCPCD) as the sensor. The PC decay curve is markedly different with a two component decay. The initial decay rate is 240 ns, followed by a long term decay of 4.72 μs. The long decay time is produced by discharge of the open-circuit device through the internal forward

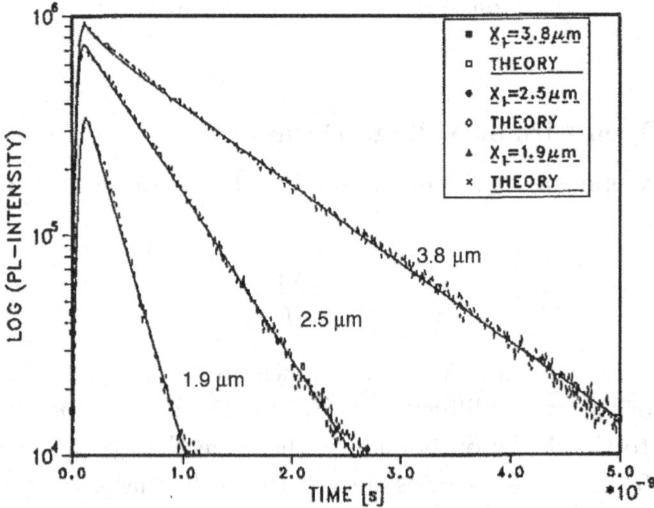

Fig. 12.6. TRPL data[5] of three structures similar doping properties but the emitter variable thicknesses shown. [Reprinted by permission from Springer Nature, *Appl. Phys.* **A53**, Transient photoluminescence decay study of minority carrier lifetime in GaAs heteroface solar cell structures, Ehrhardt, A., Wettling, W., and Bett, A. (1991) 123–129. Copyright 2019.]

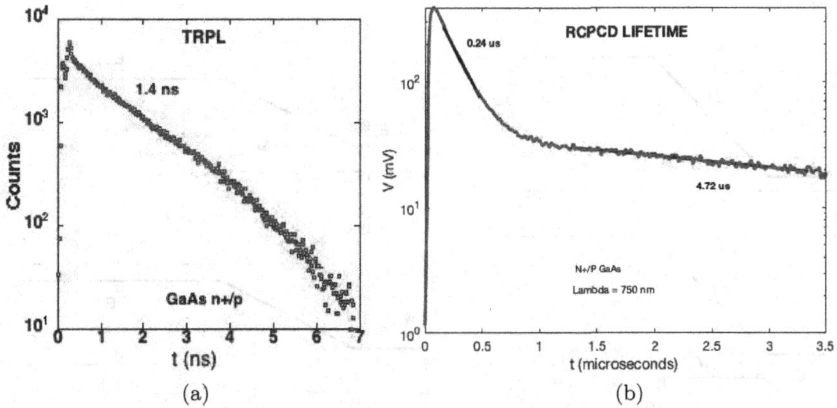

Fig. 12.7. (a) The TRPL data of a n-AlGaAs/n^+GaAs/p-GaAs/p-AlGaAs DH grown on a GaAs substrate. (b) The RCPCD decay of the same device.

bias current. The long time constant is indicative of the saturation current J_0.

12.2. Open Circuit Voltage Decay

The development of the open circuit voltage occurs in a diffusion transit time or:

$$\tau_{\mathrm{D}} = \frac{4X_{\mathrm{J}}^2}{\pi^2 D}. \tag{12.24}$$

Here, X_{J} is the base width as shown in the diagram and D is the minority carrier diffusion coefficient. Once the minority carrier diffuses to the depletion boundary, the large electric field (typically 10^4 to 10^5 V/cm) accelerates the carrier to the majority carrier side of the junction. The junction transit time is typically picoseconds in duration.

The band bending develops in the diffusion and varies from the equilibrium Fig. (12.8(a)) to and open circuit voltage Fig. (12.8(b)). The open circuit voltage (V_{oc}) decays at a much slower rate than the inherent recombination lifetime by means of back injection into the

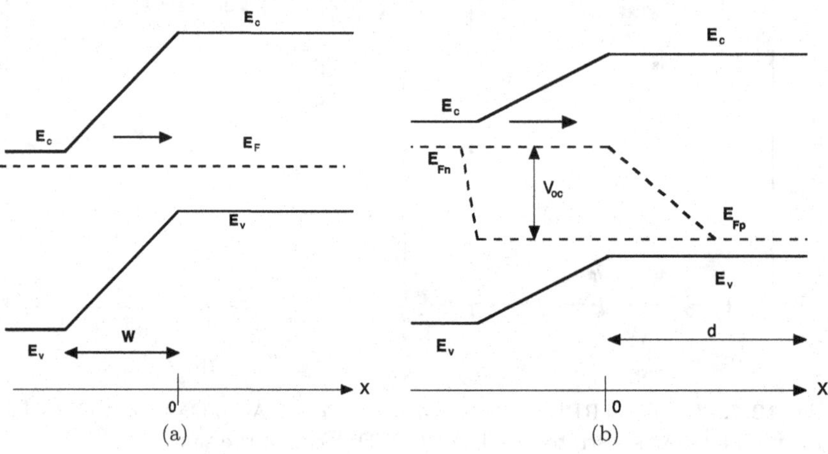

Fig. 12.8. (a) Simplified model of the band structure and electric fields in a *pn* junction at zero bias. (b) Band model of the same structure with forward bias and open circuit voltage V_{oc}.

space charge region after the pulse generation is completed. This process will be discussed later in this section.

The contacting open circuit voltage decay technique evolved many decades ago as a practical but indirect method to measure carrier lifetime in PN devices. Theory shown that the open circuit voltage of a diode decreases at a rate much more slowly than the inherent recombination lifetime. The V_{oc} decay method was accessible by the slower and less sophisticated diagnostics that were available in the 1950s era. A simple schematic diagram of the basic model is shown in Fig. 12.9. A DC bias is applied to drive a diode into forward bias and the voltage across the diode is measured in real time by an oscilloscope (or transient digitizer, today). A switch instantly turns off the bias and the voltage across the diode is monitored in real time. The decay times are analyzed and the carrier recombination lifetime of the base carrier is extracted. The model assumes that the base is sufficiently thin diffusion transit times can be neglected.

Using the band diagram in Fig. 12.6(b), the splitting of the quasi-Fermi level (or the open circuit voltage) at $t = 0$ is:

$$V_{oc} = \frac{\Delta E_f}{q} = \frac{KT}{q} \ln \frac{(n_0 + \Delta n)(p_0 + \Delta p)}{n_i^2}. \tag{12.25}$$

At low injection (Δn, $\Delta p \ll N_A$) where N_A is the base doping and $V_{oc}(0) = V_a$.

$$\frac{\partial V_{oc}(t)}{\partial t} = \frac{KT}{q} \frac{\partial}{\partial t} \ln \frac{(\Delta n)(N_A + \Delta p)}{N_A n_0}. \tag{12.26}$$

This reduces to the following in low injection:

$$\frac{\partial V_{oc}(t)}{\partial t} = \frac{KT}{q} \frac{\partial \ln(\Delta n(t))}{\partial t}. \tag{12.27}$$

Taking the electron decay in the p-type base as:

$$\Delta n(t) = \Delta n_o \exp(-t/\tau_n); \tag{12.28}$$

One calculates a linear decay of V_{oc}:

$$\frac{\partial V_{oc}(t)}{\partial t} = -\frac{KT}{q} \frac{1}{\tau_n}. \tag{12.29}$$

One may easily show that that at high injection, this approach produces a V_{oc} decay rate of:

$$\frac{\partial V_{oc}(t)}{\partial t} = -2\frac{KT}{q}\frac{1}{\tau_n}. \tag{12.30}$$

The open circuit voltage decays as KT/q times electron recombination rate, $1/\tau_n$.

Twice that rate is calculated for high injection. At 300 K, the factor KT/q is the thermal voltage or 25.8 mV. Therefore, the V_{oc} decay time is much longer than the carrier lifetime and easily measured.

Open circuit voltage decay was a standard method to measure lifetime in the early days of solid state electronics. The technique required a finished device with contacts. As other techniques evolved, it became well-known that V_{oc} decay often produced lifetime values that were inaccurate. A discussion of V_{oc} decay is presented in the text by Orton and Blood.[6] The important result here is that measurements of PN devices often do not produce accurate quantitative values of carrier lifetime. The values are generally larger than the actual lifetime because of the charge separation effect.

12.3. Contactless Measurement of V_{oc} by Transient Techniques

Currently, the recombination lifetime of PN devices can be measured by either PCD or TRPL.

The photoconductive decay technique measures the additive conductivity of both electrons and holes. Therefore, the PCD techniques measure:

$$\sigma_{pc}(t) = q(\Delta n(t)\mu_n(N_{\mathrm{D}}) + \Delta p(t)\mu_p(N_{\mathrm{A}})). \tag{12.31}$$

The changes in amplitude of the signal occur because the mobility of an electron or hole changes with the ionized impurity scattering of the region in which they are localized. For example, a minority electron in the p-type base has a larger mobility than a majority

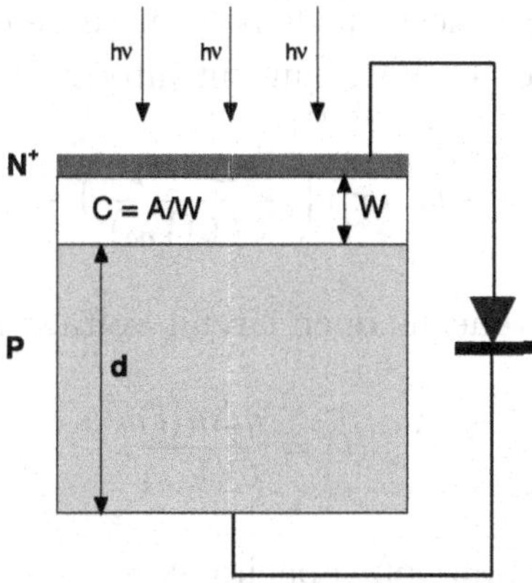

Fig. 12.9. Equivalent circuit to represent open-circuit voltage decay of a n^+/p solar cell.

electron in the n^+ emitter of an n^+/p device. We will represent the n^+/p wafer device by the equivalent circuit of Fig. 12.9.

Here, we use pulsed excitation to generated minority carriers in the n^+/p device and measure the excess carrier density decay by PCD.

First, the photoelectrons diffuse out of the base into the high electric field of the junction. Thus a transient photovoltage builds by the separation as the photoelectron density increases in the emitter. This is basic photodiode effect that underlies photodetectors and photovoltaics. However, the voltage decays (in the dark following the pulse) by injection back into the base where recombination occurs by the processes that have been discussed.

Using the model above, we compute the injection electron current into the base with open circuit voltage V_{oc}:

$$J_n = -\frac{qD_n n_0}{L_n}\left[\exp\left(\frac{qV_{oc}(t)}{KT}\right) - 1\right]. \tag{12.32}$$

Here, D_n and L_n are the electron diffusivity and diffusion length in the p-type base, respectively. The prefactor is usually called the reverse saturation current J_0.

$$J_0 \equiv \frac{qD_n n_0}{L_n}. \tag{12.33}$$

Describing the excess electron density in the emitter as $\Delta n(t)$, we can write the injected electron current into the base as:

$$q\frac{d\Delta n(t)}{dt} = -J_0\left[\exp\left(\frac{q\Delta n(t)}{KTC(V_{oc})}\right) - 1\right]. \tag{12.34}$$

Here, the instantaneous open circuit voltage is:

$$V_{oc}(t) = \frac{q\Delta n(t)}{C(V_{oc})}. \tag{12.35}$$

where $C(V_{oc})$ is the junction capacitance.

For an indirect material, such as silicon, the excess carrier density is best detected by any of the photoconductive decay techniques.

We will show that in this case, the excess electrons decay by forward bias injection and recombination in the base. The excess carrier decay time was calculated as:

$$\Delta n(t) = \Delta n_0 \exp\left(-\frac{J_0}{C}\frac{q}{KT}\right)t. \tag{12.36}$$

We can write the instantaneous lifetime as:

$$\tau(V) = \frac{KT}{q}\frac{C(V)}{J}. \tag{12.37}$$

Here, J_0 is the reverse saturation current and $C(V)$ is the capacitance of the PN junction. The capacitance, $C(V)$, is a function of junction voltage and therefore the decay process is nonexponential. As $C(V)$ is a maximum at the largest forward bias, the effective lifetime decreases as the carrier density decays. Also, the decay time decreases as J_0 increases.

In summary, the transient decay techniques produce quite different results for PCD and TRPL on the same sample. The PCD measurement produces decay times that are larger than true bulk/surface recombination lifetime. The TRPL produces data that is reflective of the minority carrier transit time and is generally shorter than the recombination lifetime. More details of these effects will be demonstrated in the next sections of this work.

12.4. Contactless V_{oc} Decay

As an example of contactless V_{oc} decay, a typical p-type silicon wafer was passivated by iodine/methanol immersion. The carrier lifetime was measured by the RCPCD technique.[7] The wafer was then processed to fabricate a n^+/p device by a shallow phosphorous diffusion into one surface of the wafer. The carrier lifetime of the n^+/p device was then measured and found to increase by more than an order of magnitude. These data are shown in Fig. 12.3. Curve A shows the RCPCD decay of the p-type silicon wafer as received from the vendor. The measured lifetime is 37 μs and is a combination of bulk and surface recombination. The wafer was etched in dilute HF and rinsed in distilled water and then immersed in iodine/methanol solution to passivate the surface. The measured lifetime is 83 μs in solution at moderate injection levels with a small decrease at low injection as shown in curve B. This value is indicative of the bulk lifetime as the surface recombination became insignificant. The wafer was removed from the solution, cleaned and then subjected to a thin phosphorous diffusion to make an n^+ layer on the p-type base. The device lifetime was then re-measured in air ambient by RCPCD and the data plotted in curve C. As shown, there is an order of magnitude increase of the initial (highest injection) lifetime to 767 μs.

The lifetime is clearly a strong function of the instantaneous injection level as predicted. Also the effective lifetime decreases continuously with time as the carrier density decreases. The final measured value is 84 μs, coincidentally about the same as the bulk lifetime. These data are consistent with the circuit model.

12.5. V_{oc} Decay Model Calculation

The dynamics of charge separation process in this experiments are calculated and the experimental results shown in Fig. 12.10. This

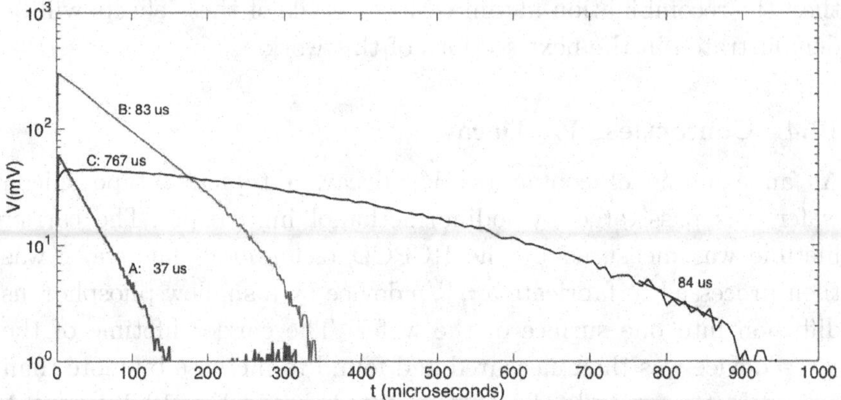

Fig. 12.10. Photoconductive decay of a n/p diode formed by phosphorous diffusion on a p-type silicon wafer. Curve A is the PCD decay prior to any treatment. The wafer was etched in dilute HF solution and immersed in iodine/methanol solution. The PCD decay was measured in-situ and the data shown in curve B. The SRV component of recombination is removed and curve B represents the bulk lifetime. Finally, the wafer is cleaned and a phosphorous diffusion produces a $n+/p$ device. The wafer is measured by PCD in air ambient and the data shown in curve C. The large increase in decay time is produced by charge separation.

calculation shows the excess electron density in the p-type base after pulsed excitation at 725 nm. This calculation requires Eq. (12.11) and uses the Fourier modal analysis terminating the infinite series after twenty terms. This calculation was performed using Matlab version 2017A and required several second of computer time. The bulk lifetime used in the calculation is that of the passivated wafer (Fig. 12.10) prior to phosphorous diffusion. The hole diffusivity is that predicted by the combination of phonon and ionized impurity scattering at a temperature of 300 K. The base is depleted of excess electrons as they flow through the left contact into the emitter. The slope at $x = 0$ is proportional to the diffusion current at each calculated time and decreases to about zero at 20 μs, indicating that open circuit voltage conditions are reached. This electron depletion shows the completion of the charge separation process. Charge separation is seen in the data (Curve C) as a finite rise in the signal near $t = 0$. The data on the bare wafer can be understood

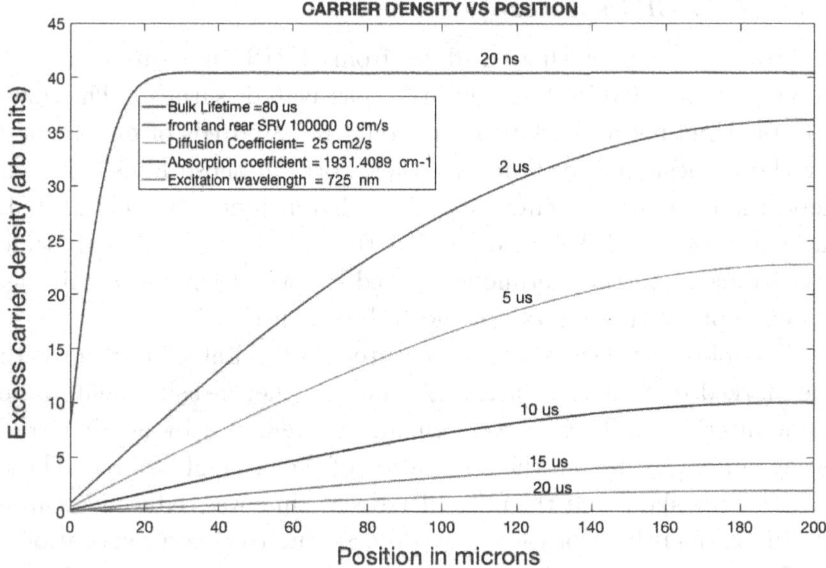

Fig. 12.11. Simulation of the charge separation process in the silicon wafer that was measured and the data shown in Fig.12.10.

in terms of the modal analysis presented earlier. Diffusion time to the surface increases the effective lifetime above that of the thin film approximation. Here, the effective thick lifetime is:

$$\begin{aligned} &\frac{1}{\tau} = \frac{1}{\tau_B} + \frac{1}{\tau_S}; \\ &where: \\ &\tau_S = \frac{d^2}{\pi^2 D} + \frac{d}{2S}. \end{aligned} \tag{12.38}$$

Using the data from curve A and B, we calculated the SRV of the unpassivated surface as about 2000 cm/s, which is a typical number for a polished silicon wafer. The SRV of the passivated wafer is too small to measure using a single incident wavelength measurement. The TRPL of silicon is also typically too weak to be measured by standard photon counting methods, but a measurement would indicate the termination of the PL transient at about 20 μs.

12.6. ZnO:CIGS Heterojunction

A previous chapter showed data from TRPL measurements on polycrystalline CIGS films with a passivated surface. The low-injection lifetime on undoped film could be hundreds of nanosecond for the highest quality films. Further work on those films[8] showed deposition on *n*-type ZnO to make a heterojunction (HJ). TRPL measured on the HJ device showed that the integrated amplitude was reduced and the lifetime decreased by two orders of magnitude. An example of those data are shown below in Fig. 12.12.

The blue curve shows data measured on the ZnO/CIGS HJ with the marked reduction in carrier lifetime. Further measurements were made after the ZnO/CdS were chemically removed by etching, and decay curve (A) showed the restoration of the original lifetime. These experiments show that the initial PL decay time was reduced because of kinetic quenching of the decay process due to charge separation.

In summary, these two transient techniques produce radically different results because of the inherent charge separation of the PN junction. These effects need to be considered in the evaluation

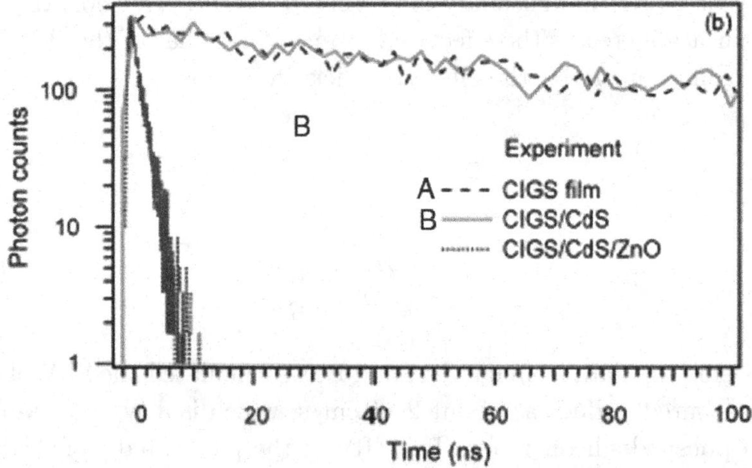

Fig. 12.12. TRPL data on bare and coated films of CIGS. The decay time is quenched with the formation of a heterojunction (blue curve). [Reprinted from Metzger, W. K., Repins, I. L., and Contreras, M.A. (2008) *Appl. Phys. Lett.* **93**, 022110, with permission of AIP Publishing.]

of experimental data. The radiative decay is proportional to the instantaneous np product, and the radiative signal goes to zero after separation at a junction. The photoconductive decay is proportional to Δn plus Δp, and a signal is produced by either majority or minority carriers. These effects have been illustrated in the three examples shown here.

12.7. Comparison of PCD and TRPL in a Confinement Structure

As a test of the theory of confinement structures and the experimental techniques, the NREL group measured the PCD and TRPL of a confinement device that had the structure n-GaInP/n-GaAs/n-GaInP. The active layer of GaAs was not intentionally doped and had a background of native donors. The photoconductive decay was measured using the RCPCD technique using the lowest possible injection. The photoluminescence decay was measured sequentially using the time-correlated single photon version of TRPL. The data are shown below in Fig. 12.13. The agreement between the two techniques is quite satisfying. Thus, a confinement structure is

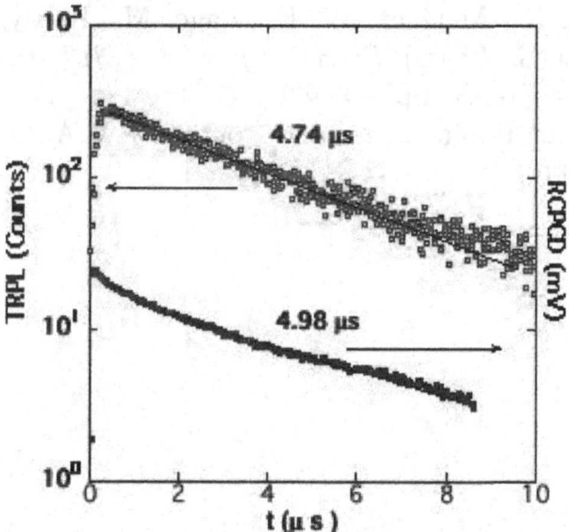

Fig. 12.13. The RCPCD and TRPL measurement of a n-GaInP/n-GaAs/n-GaInP isotype double heterostructure. As there is not charge separation here, the PCD and TRPL signals are very similar.

critical to rapid and accurate evaluation of carrier lifetime. More importantly, the correct recombination lifetime is produced by either TRPL or a PCD technique.

12.8. Summary

In summary, the charge separation process produces radically different responses to TRPL and PCD measurements. These differences must be considered in the analysis of any data.

References

1. Ahrenkiel, R. K. (1987) *J. Appl. Phys.* **62**, 2937.
2. Ahrenkiel, R. K. Dunlavy, D. J., Hamaker, H. C., Green, T. R., Lewis, C. R., Hayes, R. E., and Fardi, H. (1986) *Appl. Phys. Lett.* **49**, 725.
3. Lovejoy, M. L., Melloch, M. R., Ahrenkiel, R. K., and Lundstrom, M. S. (Invited Paper) (1992) *Solid State Electronics* **35**, 251.
4. Lovejoy, M. L., Melloch, M. R., Lundstrom, M. S., and Ahrenkiel, R. K. (1992) *Appl. Phys. Lett.* **61**, 2683.
5. Ehrhardt, A., Wettling, W., and Bett, A. (1991) *Appl. Phys.* **A53**, 123–129.
6. Orton, J. W. and Blood, P. (1990) *The Electrical Characterization of Semiconductors: Measurement of Minority Carrier Properties*, Academic Press.
7. Ahrenkiel, R. K., Metzger, W. K., Page, M., Reedy, R., Luther, J., and Dashdorj, J. (2005) *Proceedings of the 31st IEEE Photovoltaic Specialists Conference*, pp. 895–898.
8. Metzger, W. K., Repins, I. L., and Contreras, M.A. (2008) *Appl. Phys. Lett.* **93**, 022110

CHAPTER 13

Photon Recycling

13.1. Self-Absorption

Self-absorption is a direct consequence of the overlap of the absorption spectrum α(E) with the internal emission function $S(E)$. The van Roosbroeck-Shockley relationship[1] indicates that these two spectra are intimately related and must overlap. For direct band gap semiconductors, the absorption rises steeply to above $1 \times 10^4\ \text{cm}^{-1}$ for $h\nu$ greater than E_g. Therefore, the self-absorption effect is very strong in direct semiconductors. It is a much weaker effect in indirect semiconductors such as silicon. The self-absorption and re-emission of radiative recombination was proposed by Dumke[2] many years before significant experimental data were available. The term photon recycling is used to describe the generation of a new electron-hole pairs by the self-absorbed photon. The recombination event produces a new photon that may then escape or be reabsorbed in the host materials. If the photon is self-absorbed, it produces a new electron-hole pair (photon recycling) by band-to-band absorption. This process continues until the photon is externally radiated or the electron-hole pair is destroyed by nonradiative recombination.

The work of van Roosbroeck and Shockley related the emission spectrum of a semiconductor to the absorption coefficient α by thermodynamic detailed balance. This relationship between the absorption coefficient $\alpha(E)$ and the spontaneous emission spectrum

$S(E)$ was derived in which E is the photon energy.

$$S(E) = \frac{8\pi n^2 E^2 \alpha(E)}{h^3 c^2 (\exp(E/kT) - 1)} \tag{13.1}$$

In this expression, the quantities specific to a particular semiconductor are the index of refraction n, and the absorption coefficient $\alpha(E)$.

The absorption data of Fig. 13.1, curve A, from that compiled by Palik[3] and curve C is the theoretical curve calculated by applying the van Roosbroeck-Shockley expression to curve A. The data of Fig. 13.1, curve B shows the GaAs emission of a GaAs thin film measured at NREL. The GaAs film thickness is about 1.0 μm and is grown between two epitaxial layers of GaInP for surface passivation purposes. The self-absorption effect slightly distorts the emission spectra in thin devices ($d < 1\,\mu$m). One sees that the data agree well with the van Roosbroeck and Shockley calculated spectra. Because of the strong overlap between absorption and emission, self-absorption is a very prominent effect in direct semiconductors.

Figure 13.2 shows the absorption spectrum of crystalline silicon using the room temperature data of Green.[4] Also shown are the

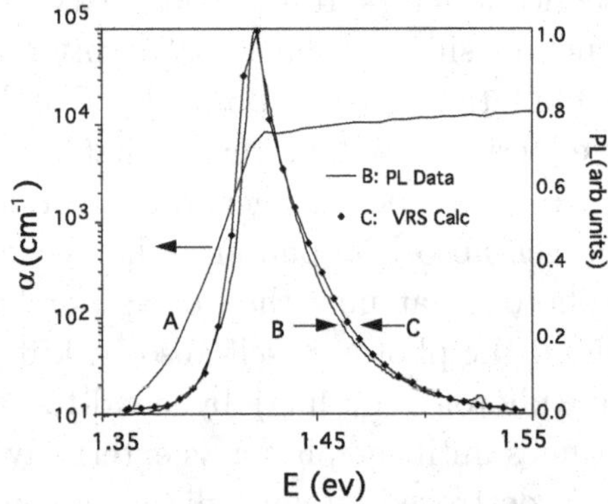

Fig. 13.1. Curve A is a plot of the absorption coefficient of undoped GaAs. Curve B is a normalized emission spectra calculated from the van Roosbroeck-Shockley relationship using Curve A and Eq. (50) with kT = 0.025 eV. Curve C is normalized PL emission data from an undoped epitaxial GaAs DH, measured at NREL.

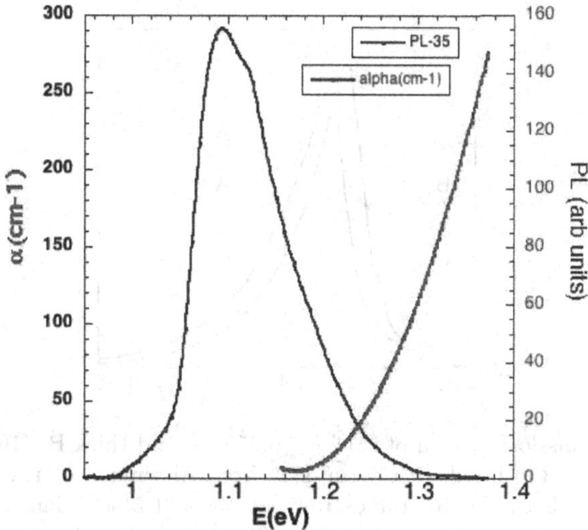

Fig. 13.2. The silicon absorption coefficients according to Green and the intrinsic emission of silicon.

emission spectra of a silicon wafer that is doped at $1 \times 10^{15}\,\mathrm{cm}^{-3}$ *p*-type that was measured at NREL. As electron-hole pairs thermalize to the bottom (top) of their respective bands, the emission occurs primarily at the indirect bandgap at about 1.1 eV. The indirect absorption edge is quite weak in the emission region and there is minimal overlap. For that reason, self-absorption in very weak in silicon and other indirect semiconductors.

13.2. Self-Absorption In Thin Films

Figure 13.3 demonstrates the strong self-absorption effects in GaAs. The emission spectra of two GaAs DH devices with different active layer thicknesses are shown here.[5] The active layer thicknesses are $0.25\,\mu$m (Curve A) and $10.0\,\mu$m (Curve B). The DH structures have the composition $Al_{0.3}Ga_{0.7}As/GaAs/Al_{0.3}Ga_{0.7}As$ and the GaAs is the doping is selenium at $1.3 \times 10^{17}\,\mathrm{cm}^{-3}$. A marked "red shift" of the emission band is seen for the "thick" $10 - \mu$m DH device of Curve B compared to the $0.25 - \mu$m device of Curve A. This red shift is produced by the stronger self-absorption of the high-energy

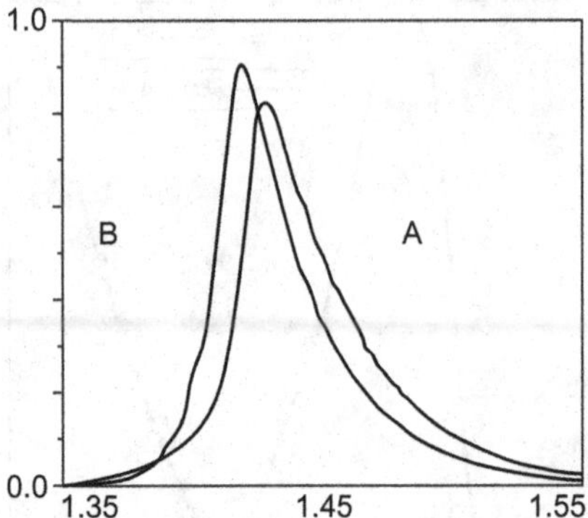

Fig. 13.3. Emission spectra of thin A, (0.25 μm) and thick B, (10.0 μm) GaAs DH Structures. [Reprinted with permission from Ahrenkiel, R. K., Keyes, B. M. Lush, G. B., Melloch, M. R., Lundstrom, M. S., and MacMillan, H. F., *J. Vac. Sci. Technol.* **A10**, 990 (1992). Copyright [2019], American Vacuum Society.]

portion of the spectrum. Self-absorption effects are observable from conventional, steady-state spectroscopy as predicted many decades ago by the calculations of Moss.[6]

A dynamic emission red shift is observed in time-resolved PL decay measurements by tuning the detecting monochromator across the PL emission band. Figure 13.4 shows data with the detector tuned to three photon energies; 1.38 eV (curve A), 1.42 eV (curve B), and 1.55 eV (curve C). The long term slopes of all decay curves are nearly identical and indicate a decay time of 465 ns. A diffusion transient is observed near $t = 0$ in curves B and C but is not in curve A. These transients are observed for all monochromator settings from about the emission peak at 1.42 eV to the high energy edge of the band at about 1.57 eV. The shapes of these TRPL curves will be compared with calculated curves to be shown later. The calculations incorporate self-absorption but not photon recycling on the transient excess carrier profiles.

A. One-dimensional self-absorption model

Significant self-absorption occurs when the diffusion length is larger than the absorption depth, $1/\alpha$, of the primary emission process.

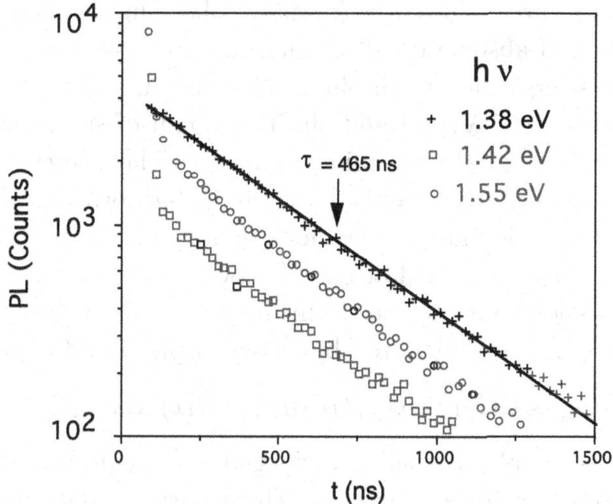

Fig. 13.4. The dynamic self-absorption effect seen in the TRPL measurement of a 10 μm DH of Fig. 12.3 by tuning across the GaAs emission band. [Reprinted with permission from Ahrenkiel, R. K., Keyes, B. M. Lush, G. B., Melloch, M. R., Lundstrom, M. S., and MacMillan, H. F., *J. Vac. Sci. Technol.* **A10**, 990 (1992). Copyright [2019], American Vacuum Society.]

The significant parameters here are:

$$L_{n,p} \gg 1/\alpha_e. \tag{13.2}$$

Here, $L_{n,p}$ is the minority-carrier diffusion length and α_e is the absorption coefficient corresponding to the observed emission wavelength.

In a previous chapter, $I_{\text{PL}}\ (t)$ was calculated by the integration of the excess carrier concentration $\rho(x,t)$ over the active volume. This calculation was carried out assuming that all of the internally generated photons are able to escape the active volume. A more exact calculation accounts for self-absorption that may prevent the escape of a large fraction of the photons. One can insert self-absorption into the modal, one-dimensional model of the transient carrier distribution. The self-absorption modifies the amplitude each of the n modes C_n. The only photons included in the calculation, are those propagated along the x-axis. This approximation is reasonably accurate for many experimental configurations in which the active photodetector area subtends a small solid angle relative to the sample. The self-absorption effect increases from the low-energy side to the high energy side of the internal emission spectrum, which is calculated from the VRS equation. The origin of self-absorption in GaAs is very clear by examining Fig. 13.1. The external emission spectra is significantly modified by the self-absorption effect as d increases. This calculation does not include the production of new photons that are generated after self-absorption.

The emission that radiates from the differential volume bounded by x and $x + \Delta x$ is attenuated by self-absorption and is given by:

$$\Delta I_{\mathrm{PL}} = \rho(x,t) \exp(-\beta(E)x)\Delta x. \tag{13.3}$$

Here $\beta(E)$ is the absorption coefficient at the monitored wavelength obtained from the absorption data. The absorption data used for the calculation came from that published by Casey and coworkers.[7] As $\beta(h\nu > E_g)$ is greater than $10^4\ \mathrm{cm}^{-1}$ over much of the emission band, self-absorption is dominant for GaAs.

The self-absorption effects account for the fast initial transient seen in time-resolved photoluminescence spectroscopy. These transients vary markedly as the detection system is tuned across the low energy side of the emission spectra. The variable parameter here is measurement wavelength (or photon energy $h\nu$) to which the detection system is tuned. Then the one-dimensional calculation is modified by the self-absorption effect as:

$$C_n(h\nu) = \int_0^d \left(\cos \alpha_n x + \frac{S}{\alpha_n D} \sin \alpha_n x \right) \exp(-\beta(h\nu)x)dx \tag{13.4}$$

The integration provides the final result as:

$$\begin{aligned} C_n(h\nu) = \frac{\beta + S/D}{\beta^2 + \alpha_n^2} - \frac{e - \beta d}{\beta^2 + \alpha_n^2} \\ \times \left\{ (\beta + S/D) \cos(\alpha_n d) + \left(\frac{S\beta}{\alpha_n D} - \alpha_n \right) \sin(\alpha_n d) \right\} \end{aligned} \tag{13.5}$$

The experimental data was simulated by inserting the modified C_n values into the modal decay curve and summing all of the modes.

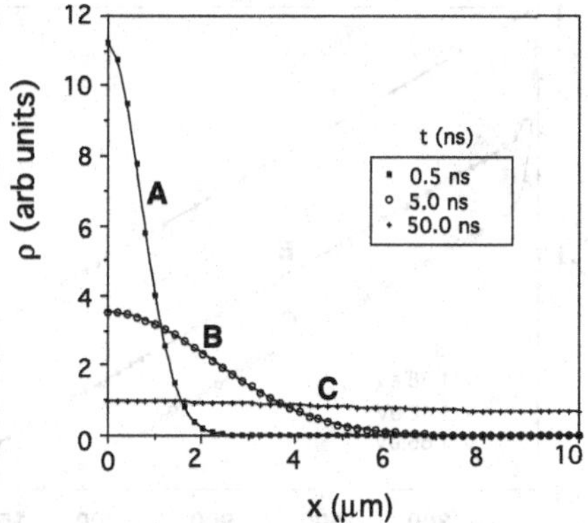

Fig. 13.5. The calculated excess carrier concentration at times of 0.5, 5, and 50 ns after the instantaneous *Beer's* law distribution.

For the profile calculations, the values of the semiconductor parameters chosen were $D = 5\,\text{cm}^2/\text{s}$, $S = 0\,\text{cm/s}$, $\tau = 500\,\text{ns}$, and $d = 10\,\mu\text{m}$.

The excess carrier profile calculation is shown in Fig. 13.5. The initial distribution has a *Beer's* law profile but the minority-carriers diffuse into the active volume very rapidly and a "flat" distribution develops. Using the modal analysis technique, the excess carrier distribution was calculated at three times after the instantaneous excitation event at $t = 0$. A "snapshot" of the function $\rho(x, t)$ is shown in Fig. 13.5 at 0.5, 5, and 50 ns after the generation of carriers at $t = 0$. The excitation absorption coefficient chosen for the calculation was $\alpha = 5 \times 10^4\,\text{cm}^{-1}$, corresponding to $\lambda \sim 600\,\text{nm}$ for GaAs. We know that at $t = 0$, the excess carriers lie in a *Beer's* law profile at the front surface. The thickness of this layer is about $1/\alpha$ or $0.5\,\mu\text{m}$ in this case. All photons generated from the surface layer escape the front surface with high probability, as self-absorption is minimal. Curve A shows the $\rho(x, t)$ distribution at $t = 0.5\,\text{ns}$, and diffusion has spread the distribution out to about $1.5\,\mu\text{m}$. Curve B is calculated with $t = 5\,\text{ns}$, and Curve C is calculated with $t = 50\,\text{ns}$. Curve C shows that the excess carrier density has become uniform by diffusion. The calculated diffusion transit time τ_D is about 20 ns. The diffusion transit time corresponds to the time between the excitation pulse and the "flat band" condition. At times $t > \tau_\text{D}$, the ρ distribution remains

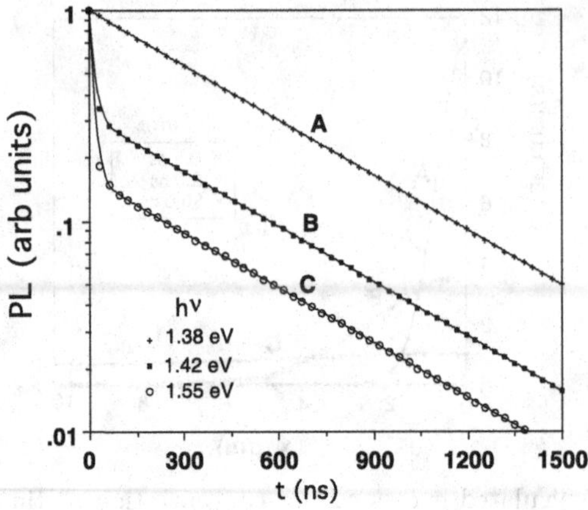

Fig. 13.6. The calculated TRPL decay curve using the modal analysis with self-absorption included.

"flat" and decreases in magnitude because of recombination. At "flat band", only the low energy photons ($h\nu < E_g$) are able to escape from the entire active volume. The only escaping higher energy photons ($h\nu < E_g$) are those generated within $1/\alpha$ from the front surface when $t > \tau_{\mathrm{D}}$. Consequently, the PL intensity drops steeply between $t = 0$ and $t > \tau_{\mathrm{D}}$ for higher energy photons ($h\nu > E_g$). Again, this calculation of $\rho(x)$ neglects the new carriers generated by the self absorption process at the calculated time.

Using these excess carrier profiles, the PL emission was computed at each time period and the self-absorption was included. For curve A, the simulation uses an emitted photon energy with hν equal to 1.38 eV, corresponding to a β-value of 12 cm^{-1} from the GaAs data. For curves B and C, the photon energies $h\nu$ are 1.42 and 1.55 eV, corresponding to β-values of 0.577×10^4 and 1.35×10^4 cm^{-1}. The self-absorption transient evolves from the calculations, as the excess carrier diffusion profile changes in time. The initial drop in intensity of higher energy photons, is produced by the increasing self-absorption as the profile "flattens" from the *Beer's* law distribution.

The calculation confirms that the origin of the initial transient is caused by the diffusion of excess carriers out of the initial *Beer's* law distribution into a uniform distribution. The behavior sometimes mimics surface recombination, so variation of absorption wavelength and monitored emission wavelengths may be required to distinguish the dominant mechanism. However, the transient TRPL data and calculations confirm the nature of the self-absorption effect.

In analyzing data for TRPL experiments, self-absorption effects my be confused as surface recombination events. More extensive measurements may be required to distinguish the principal mechanism. These include tuning both the excitation and detection wavelengths. If possible, choosing similar samples of different active layer thickness can be an importance tool in the diagnostics.

13.2.1. *Photon Recycling*

A. History

Photon recycling (PR) results from the combination of photon self-absorption and the generation of new electron-hole pairs in the device active region. When radiative recombination is dominant, then PR becomes a significant effect. The estimation and calculation of the photon recycling factor is primarily calculated by geometrical optics but can become numerically complicated. The PR effect was predicted with the calculations of Moss,[6] Kameda and Carr,[8] and Stern and Woodall.[9] The term "Photon Recycling" appears to have originated with Stern and Woodall in 1974 during their work on semiconductor lasers. They found that these self-absorbed photons lowered the laser threshold current on their semiconductor lasers. These calculations all predicted the effect was significant and would increase the effective radiative lifetime of semiconducting materials. Today, the photon recycling effect is being utilized to increase the efficiency of thin film solar cells.

The first published PR related measurements showed that $Al_xGa_{1-x}As/GaAs$ DH device lifetimes increased with the GaAs film thickness.[10] These films were doped to about $5 \times 10^{17}\,cm^{-3}$ with Te, where the predicted radiative lifetime is 10 ns. This work showed a thickness-variable lifetime that varied from about 7.5 ns at $d = 0.4\,\mu m$ to 28 ns at $d = 13\,\mu m$. There was considerable scatter in the data, probably due to SRV variation.

Observation of the PR effect was not observed until the development of a passivating confinement layer ($Al_xGa_{1-x}As$) that had a sufficiently small interface recombination velocity. Photon recycling becomes an important factor when the radiative recombination rate is larger than the surface recombination rate. Or:

$$\frac{1}{\tau_R} > \frac{2S}{d}. \tag{13.6}$$

For example, with a 1.0 μm thick active layer doped to 1×10^{17} cm^{-3}, S needs to be less than 100 cm/s in order for photon recycling to be a factor. The lifetime of the bare surface of GaAs has been estimated to be about 1×10^6 cm/s, and dominates recombination in unpassivated thin films. The epitaxial passivation process allowed radiative recombination of GaAs to be dominant, as the interface state density was reduced many orders of magnitude. Ettenberg and Kressel estimated the SRV to be between 4 and 8×10^3 cm/s in their DH devices passivated by AlGaAs. These measurements by Ettenberg and Kressel were analyzed by Asbeck[11] who suggested photon recycling was the key to the longer measured lifetimes. Asbeck also calculated the photon recycling factor, ϕ, as a function of active layer thickness and doping concentration.

Nelson and Sobers[12] reported lifetimes for a set of Al(0.5)Ga (0.5)As/GaAs DHs with doping levels from 1.9×10^{15} to 1.0×10^{19} cm^{-3}. The DH devices were grown by liquid phase epitaxy (LPE) and the lifetimes were larger than any previously reported in 1978. They measured a record (for that time) lifetime of 1.3 μs in an undoped GaAs DH as shown in Fig. 13.7.

Using the accepted value of the B-coefficient, the radiative lifetime is calculated to be 2.6 μs at 1.9×10^{15} cm^{-3} doping level. Limits can be put on the SRV based on the assumption that the bulk nonradiative recombination is negligible. Assuming that the photon recycling factor is 1.0, the SRV is;

$$\frac{2S}{d} = \frac{1}{\tau_{PL}} - \frac{1}{\tau_R}.$$

If the photon recycling factor is entered into the equation, the radiative contribution becomes negligible and the SRV controls the

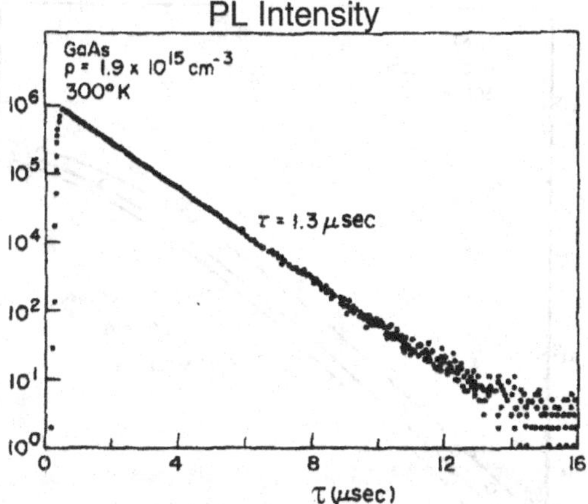

Fig. 13.7. Nelson and Sobers[12] measured a PL lifetime of 1.3 μs for the ($p = 1.9 \times 10^{15}$ cm$^-$3) doped DH structure with a thickness of 16 μm. [Reprinted from Nelson, R. J. and Sobers, R. G. *Appl. Phys. Lett.* **32**, 761, (1978), with the permission of AIP Publishing.]

PL the total lifetime:

$$\frac{2S}{d} = \frac{1}{\tau_{\mathrm{PL}}} - \frac{1}{\phi\tau_{\mathrm{R}}}.$$

This provides the limits of the SRV, as:

$$263\,\mathrm{cm/s} < S < 615\,\mathrm{cm/s}.$$

The SRV is likely the source of the dominant recombination mechanism in these measurements. A quantum efficiency measurement is needed to provide the definitive answer to the issue.

Nelson measured the lifetime versus film thickness for samples of constant doping level. These measurements were for three different doping levels. Fig. 13.8 shows data for the 5×10^{15} cm^{-3} doping level for a range of sample thicknesses from about 0.2 μm to 10 μm. The PL lifetime increases from about 20 ns at 0.2 μm to about 600 ns at the 10 μm thickness. These data were among the first to systematically show the photon recycling effect.

Later work on photon recycling effects in AlGaAs/GaAs DHs was reported by Ahrenkiel and coworkers.[14] The thin films were grown at the Spire Corporation laboratories over a range of growth

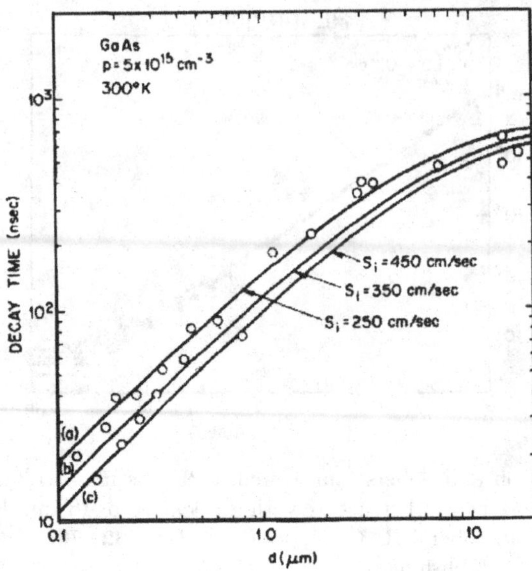

Fig. 13.8. Nelson[13] measured the lifetime versus film thickness for samples of constant doping level. [Reprinted from Nelson, R. J., *J. Vac. Sci. Technol.* **15**, 1475 (1978), with the permission of AIP Publishing.]

temperatures and growth procedures. The goals was to decrease the interface recombination velocity by optimizing growth temperature. In parallel, a photon recycling analysis was developed by K. L. Miller in 1989 at the University of Colorado and the results plotted in Fig. 13.9. A computer simulation of the companion DH devices was performed[15] using a numerical technique similar to that of Kuriyama[16] and coworkers. These calculations and those of Asbeck for the two carrier concentrations are shown in Fig. 13.9. The Miller results indicate recycling factors slightly higher than those computed by Asbeck in 1978. The calculations of Miller are shown on Curve A of Fig. 13.9. The Asbeck calculations are shown by Curve B (undoped GaAs) and curve C ($1 \times 10^{18}\,\text{cm}^{-3}$ doping).

Time-correlated single photon counting was used at NREL to measure the carrier lifetimes. The best devices had carrier decay times of 188.5 ns and 150.0 ns and were far below the predicted values of Miller and Asbeck. The data showed that the best devices (and smallest SRVs) were grown between 740°C and 800°C. The active-layer thicknesses of devices A and B are 4 and 8 μm, respectively.

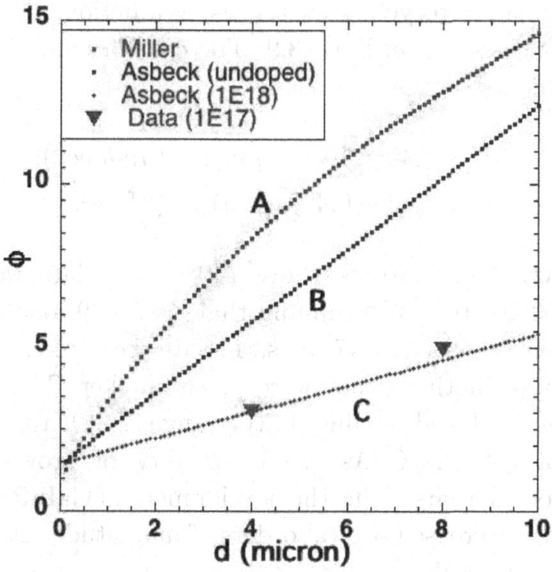

Fig. 13.9. Early calculations of the photon recycling factor for AlGaAs/GaAs DHs by K. L. Miller (curve A) and P. Asbeck (curves B and C) as a function of device thickness, d. TRPL data (reference 5) of two samples are shown for comparison. [Reprinted from Ahrenkiel, R. K., Dunlavy, D. J., Keyes, B. M., Vernon, S. M., Dixon, T. M., Tobin, S. P., Miller, K. L., and Hayes, R. E. *Appl. Phys. Lett.* **55**, 1088 (1989), with permission of AIP Publishing.]

The best devices had SRV values of $S = 216$ to $500\,\text{cm/s}$. After calculating the radiative lifetimes (using $B = 2 \times 10^{-10}\,\text{cm}^3\,\text{s}^{-1}$), the data produced photon recycling values of 3.05 (device A) and 4.94 (device B).

These measurements did show that the PL lifetimes of most of devices grown at 740°C and above are greater than the radiative lifetime. The active-layer thicknesses of devices A and B are 4 and 8 μm, respectively. The long-term decay times are 150.0 and 188.5 ns, respectively. The majority-carrier concentrations in devices A and B were measured as $1.0 \times 10^{17}\,\text{cm}^{-3}$ and $1.3 \times 10^{17}\,\text{cm}^{-3}$, respectively and the radiative lifetimes are A (50.0 ns) and B (38.5 ns), respectively. The result was that the PL lifetimes are greater than the radiative lifetimes by factors of 3.05 (device A) and 4.94 (device B). These photon PL decay data are also shown in Fig. 13.9.

The calculated recycling factor as a function of active layer thickness is also shown in Fig. 13.9. The calculations of Asbeck can be fit by a linear function.

$$\begin{aligned} \phi(d) &= 1.40 + 1.1 * d(\mu\text{m}); \quad (\textit{undoped}); \\ \phi(d) &= 1.43 + 0.4 * d(\mu\text{m}). \quad (2E18). \end{aligned} \tag{13.7}$$

For growth temperatures above 740°C, the data were fit with values of $S < 2000\,\text{cm/s}$. Assuming that $\phi = 8.39$, one calculates a recombination velocity of 447 cm/s. The devices grown at 700°C or below produced lifetimes that were much smaller. The low growth temperatures produced S values in the range 2×10^4 to 4×10^4 cm/s. The SRV of AlGaAs/GaAs as the quality of growth materials improved over the years. With the development of GaInP as a window layer, the SRV decreased several orders of magnitude relative to that of AlGaAs.

13.3. Detailed Analysis of the Recycling Process

A very detailed analysis of the recycling optics and the relationship to photovoltiacs was published by Marti[17] and coworkers in 1997. *Simultaneous* calculations by Durbin[18] and coworkers analyzed the impact of photon recycling on transient PL decay measurements. A derivation must include the primary self-absorption effects and the total internal reflection at the interface. The ray-tracing schematic of Fig. 13.10 shows some possible optical and transport events that are involved in external photon emission. We are assuming p-type doping with optical injection of electron-hole (e-h) pairs.

Process A shows an electron generated at point A_0 that diffuses to point A_1.

Using:

$$\begin{aligned} & L_n = \sqrt{D_n \tau_n}; \\ & \textit{and}: \\ & \frac{1}{\tau_n} = \frac{1}{\tau_{\text{NR}}} + \frac{1}{\tau_{\text{R}}}. \end{aligned} \tag{13.8}$$

The probability of photon emission is related to the internal QE of the event at point A1 or:

$$QE = \frac{\tau_{NR}}{\tau_{NR} + \tau_R}. \tag{13.9}$$

When electron-hole recombination occurs by radiative recombination, a photon is generated.The photon propagates through the active layer at an angle, θ, relative to the normal. The probability P_{ext} of the photon reaching the left-hand $Al_xGa_{1-x}As/GaAs$ interface is:

$$P_{ext} = \exp\left(-\frac{\alpha(h\nu)x}{\cos\theta}\right)\frac{\tau_{NR}}{\tau_{NR} + \tau_R}. \tag{13.10}$$

For indirect semiconductors such as silicon or germanium, the τ_{NR}/τ_R ratio is very small. In addition, the weak emission/absorption overlap does not allow photon recycling to be a factor for indirect materials. In Eq. (13.10), x is the distance of the generation event from the left-hand interface and θ is the angle of propagation with respect to the normal. One can define an average absorption length d_α for an emitted photon with energy $h\nu$:

$$d_\alpha = 1/\alpha(h\nu). \tag{13.11}$$

The index of refraction step (Δn) at the $Al_xGa_{1-x}As/GaAs$ interface is $n_1(Al_xGa_{1-x}As) - n_2$ (GaAs). The critical angle θ_c for total internal reflection is:

$$\theta_c = \sin^{-1}\left(\frac{n_1}{n_2}\right) \tag{13.12}$$

Reflection coefficients for transverse electrical (TE) and transverse magnetic (TM) modes have been taken into account by numerous authors. The standard approximation for the reflection coefficient $R(\theta)$, after averaging over polarization effects, is:

$$R(\theta) = 0 \quad \theta < \theta_c \tag{13.13}$$
$$R(\theta) = 1 \quad \theta > \theta_c$$

Thus, photons can only escape to the window layer when emitted within the cylindrical cone $\theta < \theta_c$. Asbeck calculated a critical angle of 72 degrees for the GaAs/Al(0.25)Ga(0.75)As interface used in the cited calculations. Thus, the probability that a photon is produced at A1 and is externally emitted is:

$$P_{ext} = \frac{\tau_{\mathrm{NR}}}{\tau_{\mathrm{NR}} + \tau_{\mathrm{R}}} \exp\left(-\frac{\alpha(h\nu)x}{\cos\theta}\right) \quad \text{for} : \theta < \theta_c. \tag{13.14}$$

Other possible paths for photon transport are indicated in the schematic in Fig. 13.10. The first possible event for photon self-absorption and reflection in a planar DH are shown in Fig. 13.10. Absorption produces the e-h pair at point A_0. The electron diffuses to point A_1. Radiative recombination produces a photon at point A_1 and the propagation angle θ is less than θ_c. The probability for

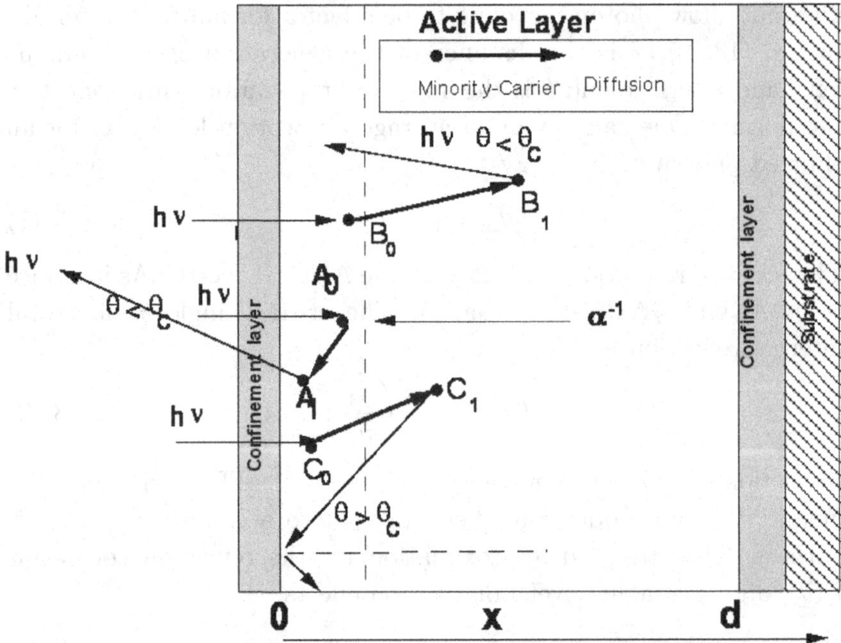

Fig. 13.10. Ray tracing of possible events in the photon emission/recycling process. Event A shows the successful escape of a eh pair generated at point A0, diffusing to point A1, and generating an escaping photon. Event B shows an eh pair generated at B0, diffusing to B1, and generating a second absorbed photon. Event C show an eh pair diffusing to C1, generating a photon that cannot escape because the emission angle exceeds the critical angle.

photon transmission into the window layer is:

$$P_{ext} = \exp(-x/L_{\alpha})(1 - R(\theta)).$$

Assuming the distance $x < L_{\alpha}$ and as $\theta < \theta_c$, the photon is able to escape from the active region.

A second transport process is indicated by the incident photon absorbed at point B_0. The photo-electron diffuses to point B_1 and by recombination, producing a second photon. The second photon is self-absorbed at B_2 by nonradiative recombination producing no photons. The energy is transferred into heat or phonons.

Thus, photon recycling is quite improbably is indirect materials because of low QE and very small overlap of emission and absorption.

A third process is indicated at point C_0. A photon is generated at an angle θ which is greater than the critical angle θ_c. A total internal reflection process occurs at the interface and reflects the photon back into the host material. This process is an important component of photon recycling in very thin films. Because a "cone" of external emission exists about the critical angle, the photon recycling factor is greater than unity, even for very thin films. The emitted photons at $\theta > \theta_C$ are reflected back into the fillm and reabsorbed in most cases. Because of the total internal reflection, the photon recycling factor is about 1.4 for films less than 1.0μm. That conclusion is supported by the calculation Fig. 13.9.

A variety of approximations have been used to calculate $G(x)$. The calculations of Asbeck and Nelson and Sobers[12] used a "flat band" approximation; i.e. $\rho(x)$ is spatially constant. Kuriyama and coworkers[16] used a technique that allows an arbitrary excess carrier profile to be used. This simulation technique was used by Miller in 1989.

After calculating $G(d)$ and including the reflection effects, one defines a quantity $F(d, \theta_{cr})$ that is the fraction of emitted photons generating electron-hole pairs by virtue of self-absorption. The lifetime can then be written as:

$$\frac{1}{\tau_{\mathrm{PL}}} = \frac{1}{\tau_{\mathrm{NR}}} + \frac{2S}{d} + \frac{1}{\tau_{\mathrm{R}}} - \frac{F(d, \theta_{cr})}{\tau_{\mathrm{R}}} \qquad (13.15)$$

Comparing Eq. (135) with Eq. (119), it is obvious that F is related to ϕ by the relationship:

$$\theta(d) = \frac{1}{1 - F(d, \theta_{cr})} \tag{13.16}$$

Thus, as $\phi(d)$ becomes very large, F approaches unity.

The time-dependent diffusion equation must be modified to include this generation effect:

$$\frac{\partial \rho(r,t)}{\partial t} = D\nabla^2 \rho(r,t) - \frac{\rho(r,t)}{\tau} + G(\rho(r)) \tag{13.17}$$

Here $G(r)$ is the generation of electron-hole pairs by self-absorption. For the one-dimensional model that has been used for DHs, Eq. (13.17) becomes:

$$\frac{d\rho(x,t)}{dt} = D\frac{d^2\rho(x,t)}{dx^2} - \frac{\rho(x,t)}{\tau} + G(\rho(x)) \tag{13.18}$$

To find $G(\rho(x))$, one must calculate the probability of self-absorption for emission from each volume element $\rho(r)\Delta V$. The emission is usually assumed to be isotropic, and therefore this problem is inherently three-dimensional. For planar structures such as DHs, cylindrical coordinates are usually used. Integration over the polar angle θ reduces the problem to a one-dimensional problem involving the cylindrical axis x. The solution also involves an integrated average over the internal emission spectrum which is convoluted with the absorption spectra α $(h\nu)$. The emission spectra are given by applying the Shockley-van Roosbroeck relationship to the given material.

Figure 13.11 shows a cylindrical coordinate system, similar to that used by Kuriyama and coworkers (1970), for calculating the photon recycling factor. For consistency with the previous one-dimensional models, the x-coordinate is the cylindrical axis that is normal to the plane of the DH film. One assumes that the film is infinite in extent in the y-z plane.

Also, the excess carrier density is uniform in the y-z plane. Then:

$$\rho(r) = \rho(r, \theta, x) = \rho(x) \tag{13.19}$$

Consider the emission from the annular ring of minority-carrier charge at point x' and radius r. The probability of a photon $h\nu$

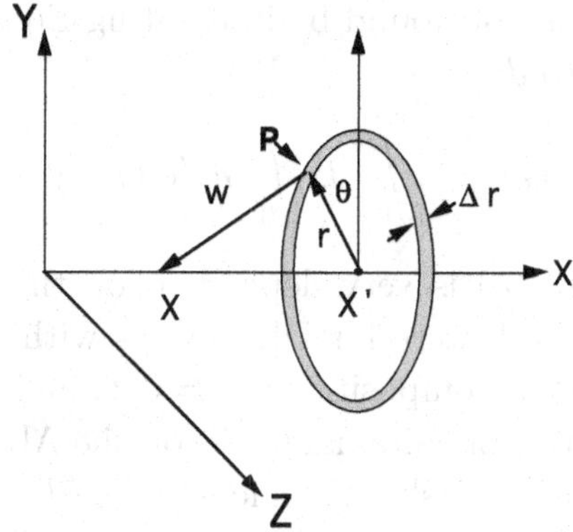

Fig. 13.11. The cylindrical coordinate system used for calculation of the photon recycling factor.

at point P (x', r, θ) reaching the cylindrical axis at point x is:

$$P = \exp(-w\alpha(h\nu)) = \exp\left(-\alpha(h\nu)\sqrt{(x'-x)^2+r^2}\right) \qquad (13.20)$$

Isotropic radiation at P is emitted into a full solid angle, and the intensity is reduced by $1/4\pi w^2$ at point x. The number of monochromatic photons of energy $h\nu$ emitted from the ring of width Δr and thickness Δx is:

$$\Delta G(x, x') = \frac{\rho(x')}{2\tau_{\mathrm{R}}} S(h\nu) \frac{r \exp\left(-\alpha(h\nu)\sqrt{(x'-x)^2+r^2}\right)}{(x'-x)^2+r^2} \Delta r \Delta x \qquad (13.21)$$

The number of monochromatic photons emitted from the semi-infinite plane at x' and thickness Δx is found by integrating r from 0 to infinity. The total photon intensity at x from the plane at x' is obtained by integrating over the normalized emission spectra $S(h\nu)$:

$$G(x, x') = \frac{\rho(x')}{2\tau_{\mathrm{R}}} \int_0^\infty d(h\nu) \int_0^\infty dr\, S(h\nu) \times \frac{r \exp\left(-\alpha(h\nu)\sqrt{(x'-x)^2+r^2}\right)}{(x'-x)^2+r^2} \Delta x \qquad (13.22)$$

The total $G(x, x')$ is now found by integrating Eq. () over the active volume or from 0 to d:

$$G(d) = \int_0^d dx \int_0^x dx' G(x', x) \tag{13.23}$$

The calculation of $\phi(d)$ is very dependent on the device structure. Calculations of Asbeck were for DH devices with $Al_xGa_{1-x}As$ confinement layers of two compositions, $x = 0.25$ and $x = 0.50$. Asbeck also added a top layer of epitaxial GaAs on the $Al_xGa_{1-x}As$ window layer is his calculation. Asbeck's calculated $\phi(d)$ for DH structures with different p- and n-type doping levels and $Al_xGa_{1-x}As$ window layer compositions are shown in Fig. 13.9. The recycling factor $\phi(d)$, calculated for the AlGaAs/GaAs DH structure, increases with the active layer thickness. These results show a linear dependence of ϕ on active layer thickness.

The calculations also show that $\phi(d)$ depends on the doping type and doping density, and decreases with the latter. These calculations predict that ϕ is greater than 10 for undoped GaAs active layers with a 10 μm thickness.

As noted earlier, as the doping level increases, the absorption edge becomes less "steep" because of the perturbation effects of ionized impurities on the band stroucture noted by Casey[7] *et al.* At values of d greater than 10 μm, $\phi(d)$ decreases rapidly for both n and p concentrations greater than 1×10^{18} cm^{-3}. At small values of d (less than 0.5 μm), $\phi(d)$ is 1.5 to 2.0 and has no doping dependence. The photon recycling factor, in this range of d, is dominated by total internal reflection. The internal reflection mechanism is verified by calculations of the lower graph of the figure. Here θ_{cr} is a parameter in the plot of $\phi(d)$ versus d. As θ_{cr} increases from 66° to 72°, $\phi(d)$ decreases from about 1.9 to 1.6. The total internal reflection disappears when the refractive indices on both sides of the boundary become identical. Looking at Fig. 13.9, we see that the PR factor at $d = 0$ is about 1.4 for both calculations. This is related to the total internal reflection effect, even at very small thicknesses.

As the effective radiative lifetime ($\phi\tau_R$) increases, the PL lifetime may become dominated by the nonradiative recombination processes.

Using Eq. (119), the PL lifetime approaches a nonradiative limit as $\phi\tau_R$ becomes very large:

$$\frac{1}{\tau_{PL}} = \frac{1}{\tau_{nR}} + \frac{2S}{d} \tag{13.24}$$

This is often the case for undoped GaAs as τ_R is 1/BN and N can be as low as $10^{14}\,cm^{-3}$ by epitaxial growth. When d is made sufficiently large, $\tau_{PL} \sim \tau_{nR}$ as SRH recombination dominates the measurement. For smaller values of d, the surface lifetime may dominate the measurement and $\tau_{PL} \sim d/2S$. The fabrication and lifetime measurement of undoped DH structures is a common method to test for the level of surface and bulk nonradiative recombination inherent in a particular growth apparatus and procedure.

A series of very detailed experiments, combined with device modeling calculations, were performed by a collaboration between NREL and Purdue University in the early 1990s. A project was undertaken to extract the radiative B-coefficient from the photon recycling factor by the NREL-Purdue group. A series of AlGaAs DHs were grown with a range of doping levels. Each doping level was represented by a range of active layer thicknesses, d. The experimental data are shown in publications by the Purdue-NREL collaboration.[5,19–21] Durbin and Gray made extensive calculations of the photon recycling factor in the experimental structures.[22]

Some data from the above project are shown in Fig. 13.12 at four different doping densities. As the direct bandgap absorption edge becomes steeper with decreasing doping density, the self-absorption also decrease with increasing doping density. DH structures were made at four different doping levels over a range of n-type doping from 1.3×10^{17} to $2.2 \times 10^{18}\,cm^{-3}$ and some of these data are shown below.[21] The decrease in the photon recycling factor with increased doping is very obvious in these data.

Figure 13.13 plots the PL lifetime of 0.25–0.50 μm and 8–10 μm DH structures from each doping concentration. The solid line is a plot of the radiative lifetime again assuming $B = 2 \times 10^{-10}\,cm^3/s$. All of the data points fall on or above the radiative lifetime. The data points labeled A are from devices either 0.25 μm or 0.50-μm

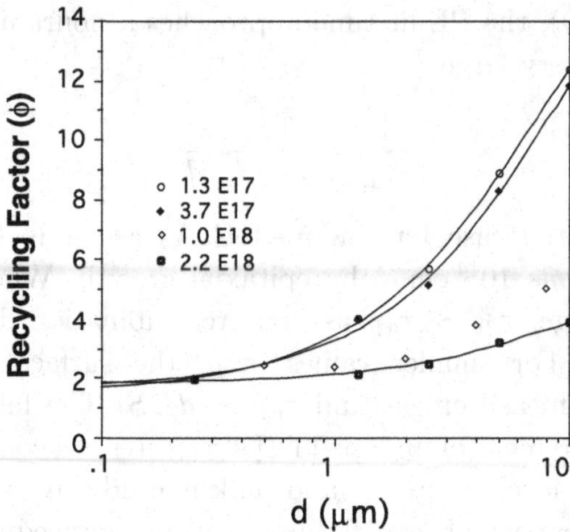

Fig. 13.12. This a plot of measured photon recycling factors using the S-value from the thinnest film. There were four n-type concentrations used and range of film thickness for each concentration. The solid curves are the calculated predictions of Asbeck in 1977 are those of equation (13.7). The data are from reference 21. [Reprinted from Lush, G. B., Melloch, M. R., Lundstrom, M. S., Levi, D. H., Ahrenkiel, R. K., and Macmillan, *Appl. Phys. Lett.* **61**, 2440 (1992), with permission of AIP Publishing.]

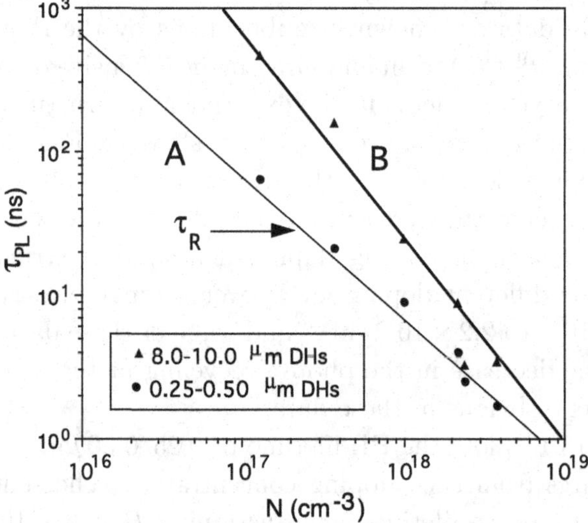

Fig. 13.13. Curve A is a plot of predicted lifetime with $B = 2 \times 10^{-10}\ \text{cm}^{-3}\text{s}^{-1}$. As the samples are thin, photon recycling is minimal for these data. Curve B is a plot of $\tau_{\text{PL}} = 12.2 \times \tau_{\text{R}}$ because of photon recycling in the thicker samples.

thick, and the points generally lie slightly higher than τ_R. The thick devices labeled B are either 8.0-μm or 10.0-μm thick. The PL lifetime of the 10-μm device from the $1.3 \times 10^{17}\,\text{cm}^{-3}$ series is 12.2 times larger than the radiative lifetime. As the doping density increases, τ_{PL}/τ_R decreases. This decrease is a result of the drop in $\phi(d)$ with increased doping and will be discussed later. The A data points are representative of the radiative lifetime and agree quite well with the accepted B-value.

The radiative lifetimes are consistent with earlier data (Nelson, Nelson and Sobers) and theory with $B = 2 \times 10^{-10}\,\text{cm}^{-3}\text{s}^{-1}$.

Garbuzov and coworkers[23] calculated $\phi(d)$ for a "free standing" DH, i.e. one for which the substrate had been etched away. These calculations indicated a large increase in $\phi(d)$ as light trapping occurs at both interfaces in this case. Later calculations by Durbin and coworkers expanded on the substrate removal effect.

A markedly large increase of the photon recycling effect was reported by etching away the substrate on a GaAs DH structure by Lush[21] and coworkers in 1992. This process increases the total internal reflection effect, by decreasing the critical angle. Thus, only photons near normal incidence can avoid total internal reflection. A hole was etched through the substrate of the DH structures described earlier. The TRPL measurements were made in regions with the substrate removed and compared with the unetched regions. The data for the thickness series with doping of $N = 1.3 \times 10^{17}\,\text{cm}^{-3}$ are shown in Fig. 13.14. The radiative lifetime calculated with a B-coefficient of $2 \times 10^{-10}\,\text{cm}^3\text{s}^{-1}$ is 38 ns. For the unetched portion of the device, one sees that the PL decay time reaches about 380 ns or ten times the radiative lifetime for the 10-μm thick sample. However, with the substrate removed, the maximum lifetime is 1070 ns or twenty eight times the radiative lifetime.

This result verifies earlier calculations that photon recycling is greatly enhanced by substrate removal. As shown by Fig. 13.14, the PL lifetime varies from about 500 ns for thin DHs to over 1 μs for thicker DHs. These data suggest that very long lifetimes are feasible in more heavily doped, free-standing thin-film devices because of the photon recycling effect. This dramatic effect was predicted a decade earlier

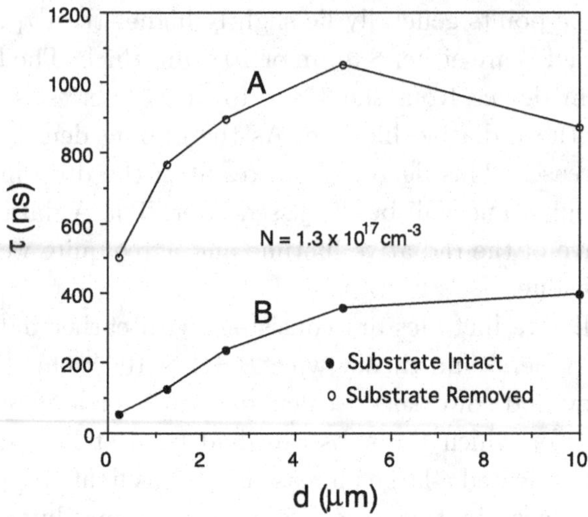

Fig. 13.14. The data of Lush[21] and coworkers for lifetime measurements on a thickness series of AlGaAs/GaAs DH devices all doped to $1.3 \times 10^{17}\,\text{cm}^{-3}$. Curve B is the data before substrate removal and Curve A is the same sample after removal. [Reprinted from Lush, G. B., Melloch, M. R., Lundstrom, M. S., Levi, D. H., Ahrenkiel, R. K., and Macmillan, *Appl. Phys. Lett.* **61**, 2440 (1992), with permission of AIP Publishing.]

by Garbuzov *et al.*[23] As shown by Fig. 13.13, the PL lifetime varies from about 500 ns for thin DHs to over 1 μs for thicker DHs. These data show that very long lifetimes are feasible in more heavily doped, free-standing thin-film devices because of the photon recycling effect.

A classic study[24] produced a record lifetime for an undoped GaInP/GaAs DH grown by MOCVD. The DH was grown on a GaAs substrate with a two thin GaAs buffer layers grown on the substrate, followed by the 1.0 μm thick GaAs active layer. The window layers were composed of lattice-matched GaInP and were 30 nm thick. The growth temperature for the GaAs layer was 625°C and for the GaInP window layers was 575°C.

Several devices were grown with PC decay times exceeding 10 μs. Data from the "champion" sample are shown in Fig. 13.14 and had low injection lifetimes of 120 to 130 μs. The PC decay was measured by the RCPCD technique and the injection level was varied over a wide range. Using the B-value of $2 \times 10^{-10}\,\text{cm}^3\,\text{s}^{-1}$ and a photon recycling factor of 2.5 (from equation (13.7), the upper limit

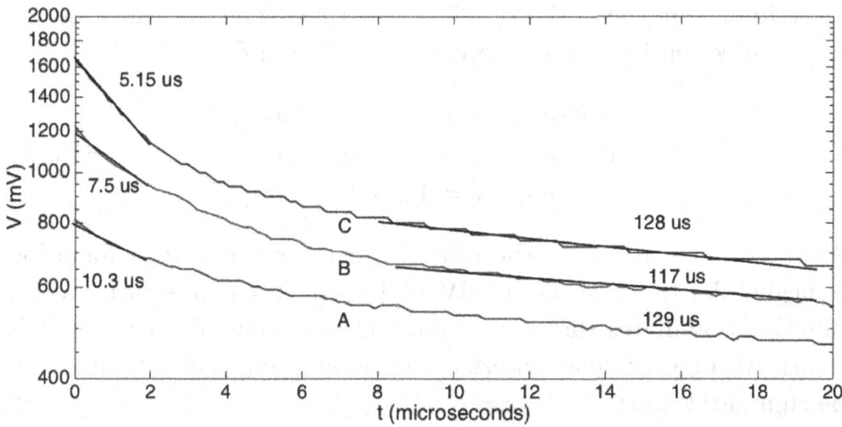

Fig. 13.15. Photoconductive decay (RCPCD) on an undoped GaInP/GaAs DH with and active layer thickness of 1.0 μm. The measurements were made a three injection levels as indicated by curves A, B, and C. This work is the unpublished data of S.P. Ahrenkiel.

on background doping (assuming SRV $= 0$) is $1.04 \times 10^{14}\,\text{cm}^{-3}$. The other limit is that the decay is SRV limited. In that case, $S = 0.41\,\text{cm/s}$.

Assuming that the initial decay is determined by bimolecular high radiative recombination, one can use the initial decay time to calibrate the injection level. Using the previously derived result for high radiative injection:

$$\begin{aligned} \sigma(t) &= q * (\mu_n + \mu_p)B(\rho N + \rho^2); \\ \rho &\equiv \Delta n = \Delta p. \end{aligned} \tag{13.25}$$

Here, N is the background doping, either n or p and we assume that the recombination is purely radiative at high injection.
At $t = 0$, and $\rho \gg N$, one can write:

$$\frac{\left(\frac{d\sigma}{dt}\right)}{\sigma} = 2B\rho/\phi = \frac{1}{\tau}; \tag{13.26}$$

Therefore, from the initial ($t = 0$) decay times, one can calculate the initial injection levels for data curves A, B, and C.

$$\begin{aligned} &\text{Curve A: } \rho_0 = 9.7 \times 10^{14}\,\text{cm}^{-3} \\ &\text{Curve B: } \rho_0 = 1.3 \times 10^{15}\,\text{cm}^{-3} \\ &\text{Curve C: } \rho_0 = 1.9 \times 10^{15}\,\text{cm}^{-3} \end{aligned} \tag{13.27}$$

At the other extreme, if the radiative component of recombination is negligible, the maximum SRV is 0.41 cm/s and interface recombination dominates the asymptotic lifetime. Thus, the net result is indicative of very small interface recombination. These results can be summarized as:

$$\begin{aligned} &\frac{1}{\tau_{\text{PL}}} = BN + \frac{2S}{d}; \\ &N \leq 1.04 \times 10^{14}, \text{cm}^{-3}; \\ &S \leq 0.41, \text{cm/s}. \end{aligned} \tag{13.28}$$

13.3.1. *Device Applications*

The photon recycling effect has been utilized in recent years to make more efficient thin film solar cells. Early work by Lush and Lundstrom[25] suggested photon recycling for improved photovoltaic cells as well as other devices. Durban and Gray[26] used a computer model to calculate the effects of photon recycling on a GaAs solar cell. Their results predict an increase in the open circuit voltage of 63 mV for a thin film cell that has been lifted off of the substrate.

In 1990, Tobin[27] and coworkers reported a design with a predicted open-circuit voltage of 1.029 volts and a one-sun efficiency of 24.8%. In recent years, the open-circuit voltage enhancement by photon recycling has become very common in thin-film photovoltaic devices. A record efficiency of 27.6% was measured by Kayes and coworkers[28] for a GaAs one sun, single junction device that was fabricated by epitaxial lift-off (ELO). The open-circuit voltage was increased to 1.107 volts, which is an increase of 78 mV relative to the device of Tobin and all. The back surface was a metal film that increased the photon recycling effect over that of total internal reflection.

Advanced designs, based on photon recycling, have been analyzed by a number of groups for designing high efficiency GaAs-based solar cells. These included Lumb[29] and coworkers, Miller and coworkers,[30] and Kosten and coworkers.[31] Kosten and all used an external dielectric angle restrictor adjacent to a device and saw a 3.6 mV increase in the open circuit voltage. Most of these devices utilize the suppression of the forward-bias or "dark current" that is reduced by the photon recycling effect.

Watanabe and coworkers[32] analyzed the use of photon recycling in a quantum well solar cell to increase the performance. Xiao[33] and coworkers proposed a thermophotovoltaic device with photon recycling that would make substantial increases in the conversion efficiency in a In(0.53)Ga(0.47)As based cell for use in a 1200°C device.

Photon recycling has proven to be a very useful effect over many decades. When incorporated into diode lasers, the PR effect lowered the lasing threshold and increased the efficiency of these devices. Currently, the PR effect is being incorporated into the design of higher efficiency thin film photovoltaic devices. In summary, the PR mechanism continues to find new applications and to find a place in new device designs.

References

1. van Roosbroeck, W. and Shockley, W. *Phys. Rev.* **94**, 1558, (1954).
2. Dumke, W. P., *Phys. Rev.* **105**, 139 (1957).
3. Palik, E. D. (1985). In *Handbook of Optical Constants of Solids*, Academic Press, Orlando, Florida.
4. Green, M. A. "Silicon Solar Cells", Centre for Photovoltaic Devices and Systems, University of New South Wales, (1995).
5. Ahrenkiel, R. K., Keyes, B. M.. Lush, G. B., Melloch, M. R., Lundstrom, M. S., and MacMillan, H. F., *J. Vac. Sci. Technol.* **A10**, 990. (1992).
6. Moss, T. S. *Proc. Phys. Soc.* (London), **B70**, 247, (1957).
7. Casey, H. C., Jr., Sell, D. D., and Wecht, K. W. *J. Appl. Phys.* **46**, 250. (1975).
8. Kameda, S. and Carr, W. N. *J. Appl. Phys.* **44**, 2910, (1973).
9. Stern, F. and Woodall, J. M., *J. Appl. Phys.* **45**, 3904, (1974).

10. Ettenberg, M. and Kressel, H. *J. Appl. Phys.* **47**, 1538, (1976).
11. Asbeck, P. *J. Appl. Phys.* **48**, 820. (1977).
12. Nelson, R. J. and Sobers, R. G. *Appl. Phys. Lett.* **32**, 761, (1978).
13. Nelson, R. J., *J. Vac. Sci. Technol.* **15**, 1475 (1978).
14. Ahrenkiel, R. K., Dunlavy, D. J., Keyes, B. M., Vernon, S. M., Dixon, T. M., Tobin, S. P., Miller, K. L., and Hayes, R. E. *Appl. Phys. Lett.* **55**, 1088. (1989).
15. Miller, K. L. "Analysis of Transient Photoluminescence from $Al_xGa_{1-x}As$/GaAs Double Heterostructure Samples," M.S.E.E. Thesis, University of Colorado at Boulder (1989).
16. Kuriyama, T., Kamiya, T., and Yanai, H. *Jap. J. Appl. Phys.* **16**, 465. (1977).
17. Marti, A., Balenzategui, J. L., and Reyna, R. F. *J. Appl. Phys.* **82**, 4067 (1987).
18. Durbin, S. M., Gray, J. L., Ahrenkiel, R. K., and Levi, D. H. "Numerical Modeling of the Influence of Photon Recycling on Lifetime Measurements", 23nd IEEE Photovoltaics Specialists Conference-1993, p. 628.
19. Lush, G. B., MacMillan, H. F., Keyes, B. M., Ahrenkiel, R. K., Melloch, M. R., and Lundstrom, M. S. In *Proceedings of the Twenty-Second IEEE Photovoltaic Specialists Conference, Las Vegas, Nevada*, p. 182. IEEE, New York. (1991).
20. Lush, G. B., MacMillan, H. F., Keyes, B. M., Ahrenkiel, R. K., Melloch, M. R., and Lundstrom, M. S. *J. Appl. Phys.* **72**, 1436, (1992).
21. Lush, G. B., Melloch, M. R., Lundstrom, M. S., Levi, D. H., Ahrenkiel, R. K., and Macmillan, *Appl. Phys. Lett.* **61**, 2440 (1992).
22. S. M. Durbin and J. L. Gray, in *Proceedings of the 22nd IEEE Photovoltaics Specialists Conference* (Las Vegas, NV October 1991), 1991, p. 188.
23. Garbuzov, D. Z., Ermakova, A. N., Rumyantsev, V. D., Trukan, M. K., and Khalfin, V. B. *Sov. Phys. Semicond.* **11**, 419 (1977).
24. S. P. Ahrenkiel (unpublished data, 2018).
25. Lush, G. B. and Lundstrom, M. S. (1991). *Solar Cells* **30**, 337.
26. Durbin, S. M. and Gray, J. L. *IEEE Trans. Electron Dev.* 41, 239–245 (1994).
27. Tobin, S. P., Vernon, S. M., Bajgar, C., Wojtczuk, S. J., Melloch, M. R., Keshavarzi, A., Stellwas, T. B., Venkatensan, S., Lundstrom, M. S., and Emery, K. A. *IEEE Trans. Electron Devices*, **37**, 469 (1990).
28. Kayes, B. M., Nie, H., Twist, R., Spruytte, S. G., Reinhardt, F., Kizilyalli, I. C., and Higashi, G. S. in 37th IEEE Photovoltaic

Specialists Conference (IEEE, Seattle, WA, 2011), p. 4.

29. Lumb, M. P., Steiner, M. A., Geisz, J. F., and Walters, R. J. *J. Appl. Phys.* **116**, 194504 (2014).
30. Miller, O. D., Yablonovitch, E., and Kurtz, S. R. *IEEE Journal of Photovoltiacs2*, 303–311 (2012).
31. Kosten, E. D., Kayes, B. M., and Atwater, H. A. *Energy Environ. Sci.* **7**, 1907–1912 (2014).
32. Watanabe K, Inoue T, Toprasertpong K, Delamarre A, Sodabanlu H, Guillemoles JF, Sugiyama M, Nakano Y. 2016 IEEE 43rd Photovoltaic Specialists Conference (PVSC). pp. 1268–1272, (2016).
33. Xiao TP, Scranton G, Ganapati V, Holzrichter J, Peterson P, Yablonovitch E. 2016. Enhancing the efficiency of thermophotovoltaics with photon recycling. In: 2016 Conference on Lasers and Electro-Optics (CLEO). pp. 1–2.

CHAPTER 14

Simultaneous and Comparative Measurements

The emphasis of this work has been the measurement of the minority-carrier lifetime of photovoltaic materials. Carrier lifetime is a high-level, critical parameter for the operation of photovoltaic devices. There are inherent artifacts that are inherent in the measurement techniques and these often mask the true excess carrier lifetime. Here, we will discuss sequential and simultaneous measurements in an attempt to compare measurement techniques directly. The time-dependent transient techniques contain a great deal of information about the important physical processes that are contained in the functional form of the signal decay. The time dependent photo-conductive decay process measures $\Delta\sigma(t)$, and $\Delta n(t)$ is extracted from the data. The conductivity decay is the product of $\Delta n(t)$ and mobility, $\mu(\Delta n)$. For analysis, the mobility is often assumed to be constant over the measured range of $\Delta n(t)$. However, if the excess carrier density can be estimated, the mobility correction can be incorporated in the analysis software. In comparing PCD and TRPL, theory shows a high-injection variation between PCD and TRPL decay functions. The PL decay rate varies as $N\Delta n(t)+\Delta n(t)^2$ where N is the background doping. This bimolecular decay is an identifier of dominant radiative decay. Most of these effects can be viewed as a source of additional information, rather than a source of conflict between measurements. Photoconductive decay measures the conductivity of all excess carriers (both majority and minority carriers), whereas photoluminescence decay measures only direct electron-hole recombination via photon emission. As radiative recombination is the dominant mechanism in the tandem cells used for concentrator technology, these high injection effects are especially important for this technology.

When a shallow trap captures a minority carrier, the instantaneous photoluminescence signal is terminated. However, the associated majority carrier, created during electron-hole pair generation, continues to provide a photocurrent. The majority carrier signal persists until a new recombination event occurs with another free minority carrier. The minority carrier may be generated by emission from a shallow trap and be delayed by the thermal emission time. Often the competition between recombination and shallow traps leads to complications in data interpretation.

14.1. Simultaneous PCD and TRPL Measurements

We will examine several comparison studies of the same material by different techniques. The challenging aspect of these studies is the dependence of recombination rate on illumination intensity (i.e. injection level) and the complication of moving the sample to different measurement systems. A more recent innovation is the development of simultaneous measurement techniques in which two or more measurements are simultaneously made. One example here is the simultaneous measurement of PCD and PL decay. Both dual and triple simultaneous measurements will be described here.

14.2. Dominant Radiative Recombination

The dominant recombination mechanism in direct bandgap, epitaxial, III-V and IV-VI materials is usually radiative in nature. PCD and TRPL are the most common methods to measure carrier lifetime in materials. Here, we will explore differences that one might expect, in these two measurements. The TRPL signal has been given by:

$$I_{\mathrm{PL}}(t) = B^*(\rho N + \rho^2); \tag{14.1}$$

where: $\rho = \Delta n$ or Δp.

The observable lifetime, τ_{PL}, in a PL decay measurement is:

$$\frac{1}{\tau_{\mathrm{PL}}} = -\frac{dI_{\mathrm{PL}}(t)/dt}{I_{\mathrm{PL}}(t)} = -\frac{N+2\rho}{N+\rho}\left(\frac{1}{\rho}\frac{d\rho}{dt}\right); \tag{14.2}$$

If one is measuring the photoconductive decay simultaneously, the result is;

$$\frac{d\sigma(t)}{dt} = q(\mu_n + \mu_p)\frac{d\rho}{dt}; \tag{14.3}$$

The photoconductive decay time, τ_{PCD}, is:

$$\frac{1}{\tau_{\mathrm{PCD}}} = -\frac{1}{\rho}\frac{d\rho}{dt}; \tag{14.4}$$

Therefore, the relationship between the two measurements is:

$$\frac{\tau_{\mathrm{PCD}}}{\tau_{\mathrm{TRPL}}} = \frac{N+2\rho}{N+\rho}. \tag{14.5}$$

Therefore, at low injection, the measured values are equal. At high injection ($\rho > N$), the photoconductive lifetime is two times the photoluminescence lifetime. This effect has been measured and will be demonstrated later in this chapter.

The effect may be easily simulated for the case of GaAs with a known B-coefficient. In Fig. 14.1, the simulation assumes GaAs that is doped to a level of $1 \times 10^{16}\,\mathrm{cm}^{-3}$. The model assumes no SRH recombination and a pulse produces $2 \times 10^{16}\,\mathrm{cm}^{-3}$ carriers that are uniformly injected into the two micron-thick GaAs film. The low injection lifetime is $0.5\,\mu\mathrm{s}$ and the high injection effects are apparent from the model. The photoconductivity is proportional to the excess carrier density, $\rho(t)$. At the initial stage of the decay, the decay is nonexponential as the $\rho^2(t)$ term dominates in the radiative recombination model. However, the PL decays decreases at a faster rate in accord with the theory above. At low injection, the two rates become equal.

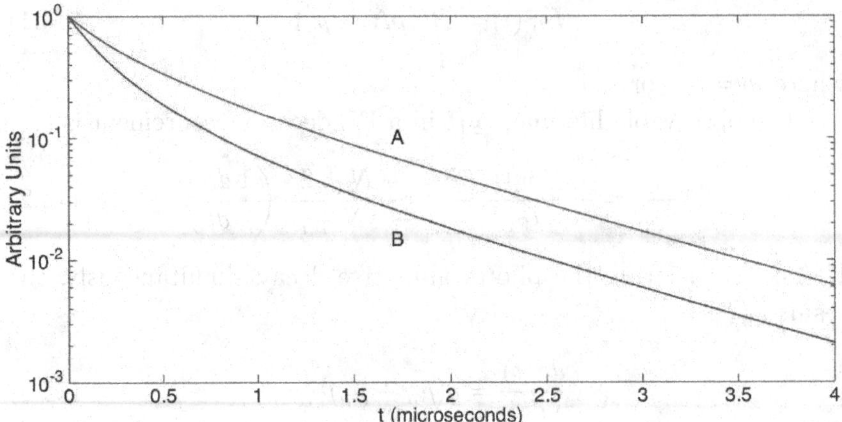

Fig. 14.1. Calculated signal decay of a 2 μm thick GaAs DH doped to 1×10^{16} cm^{-3}. The initial injection pulse deposits an excess carrier density of 2×10^{16} cm^{-3} electron-hole pairs. Curve A is the photoconductive decay assuming constant mobility. Curve B is the photoluminescence decay. [© (2019) IEEE. Reprinted with permission, from Ahrenkiel, R. K., Feldman, A., Lehman, J., and Johnston, S. W. *IEEE J. Photovoltaics* **3**, pp. 348–352 (2013).]

Experimental studies of a similar structure will be shown later here. At low injection, the decay curves become identical.

14.3. Comparison of PCD Measurements of Multi-crystalline Silicon

Multi-crystalline silicon has become the dominant materials for the low-cost, residential photovoltaic application. Because of the grain boundary issue, lifetime characterization is somewhat more complicated.

A commercial, multi-crystalline silicon wafer was masked and measured by μPCD mapping, and RCPCD and QSSPC for comparison purposes.[1] The sample was masked into four equal-sized areas and measured by two other PCD techniques after removal from the QSSPC apparatus. The QSSPC lifetime versus injection level data were plotted earlier in Fig. 9.7 to illustrate the longer lifetimes at the low injection levels. This effect was linked to by shallow traps at grain boundaries. At excess carrier densities of between 3- and $5 \times 10^{14}\,\text{cm}^{-3}$, the QSSPC lifetime saturates at a value that corresponds to the recombination lifetime. These lifetimes were compared to measurements on the masked areas by both the Semilab μPCD scanning system and the NREL RCPCD system. The μPCD data were measured on the masked areas with the Semilab WT-2000PV scanning system. The lifetime map, Fig. 14.2(a) shows the large spatial variation of the carrier lifetime within each masked area, as well as variations between the separate area.

For the RCPCD measurement, each masked are A, B, D, and E is measured sequentially. In this particular measurement, the laser beam was defocused to a diameter; slightly larger than the 1.5-inch aperture of the mask. We then measured the transient decay data in these four different exposed regions of the wafer.

These data are shown in Fig. 14.2(a). The incident wavelength of the excitation pulse was 900 nm for the RCPCD measurements.

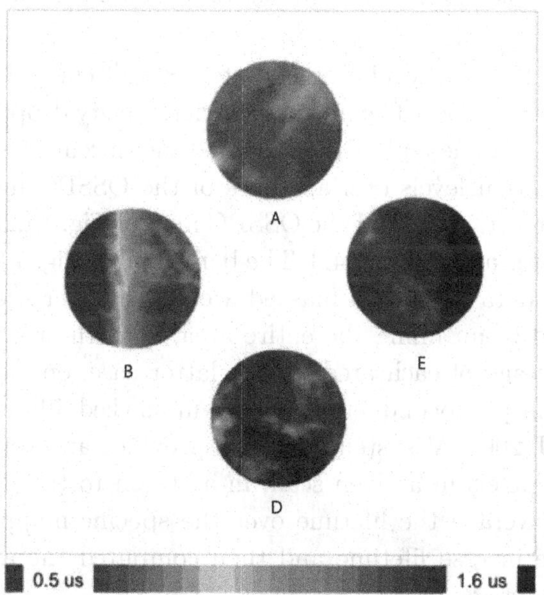

Fig. 14.2(a). The μPCD map of masked areas on a multicrystalline silicon wafer using the Semilab WT-2000PV lifetime scanner. The lifetimes are color coded and the scale is shown in the bottom color bar. A large defective area lines the left edge of the wafer that appears in area B.

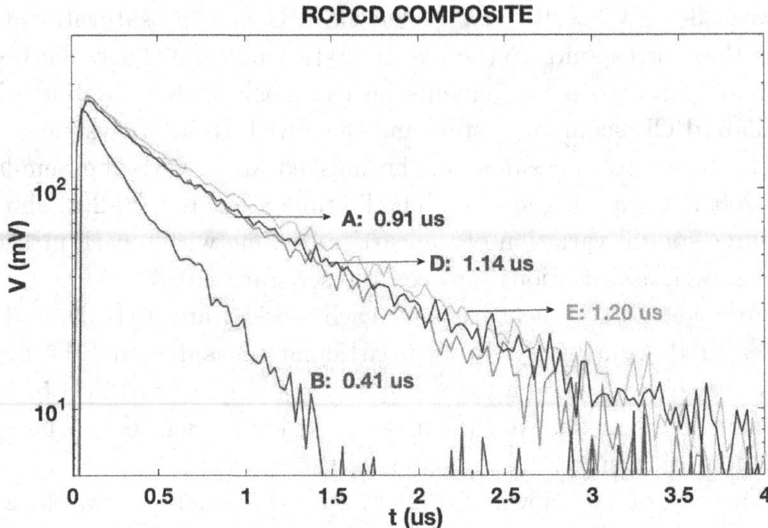

Fig. 14.2(b). The RCPCD measurement of the same masked areas that are shown in the lifetime map. Area B contains the defective volume.

The initial injection level was set at $1 \times 10^{16}\,\text{cm}^{-3}$ by means of calibrated attenuators. The injected carrier density dropped to about $2 \times 10^{14}\,\text{cm}^{-3}$ at the end of the recorded measurement. Thus, these RCPCD injection levels overlap those of the QSSPC measurement. The higher injection data of the QSSPC map of Fig. 14.2(a) are plotted in the bar graph of Fig. 14.3. The bar graph of Fig. 14.3 also plots the RCPCD data from each masked area. The latter are the results of the pulsed beam filling the entire area, with the resultant signal being an average of each area. In the latter case, one measures the total transient photoconductivity of the unmasked, illuminated area.

This WT-2000PV system software provides an average lifetime of all of the pixels in a given scan, in addition to the lifetime map. In order to average the lifetime over the specific mapped area, we averaged the inverse lifetime and then computed the average from the inverse average.

$$\frac{1}{\tau_{ave}} = \frac{1}{N}\sum_{i=1}^{N}\frac{1}{\tau_i} \tag{14.6}$$

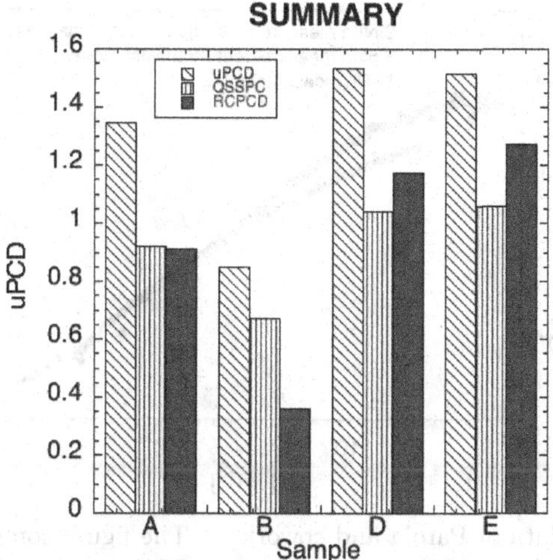

Fig. 14.3. A chart of each masked area of the multi-crystalline wafer with a value provided by each of the three measurements; μPCD, QSSPC, and RCPCD.

As the lifetime is the inverse of the recombination probability, the probabilities are additive as shown earlier. That average for each masked area is entered in the barograph of Fig. 14.3.

Agreement between the QSSPC and RCPCD data are quite good considering the very different measurement methods. The data from area B, with the lowlifetime region, shows a PCD lifetime that is about one-half of the average inverse map-lifetime. There is considerable variation among techniques for sample B that has the large low-lifetime region. These results show that all of these techniques give an acceptable average for an area of multi-crystalline silicon. There is more variation with the computed μPCD average than with the directly measured averages. In addition, the μPCD map has the added advantage of giving detailed spatial information about defective areas. The disadvantage of mapping is a somewhat longer data acquisition time, then either QSSPC or RCPCD.

Parola and coworkers[2] did a series of measurements on lightly doped n-type silicon (3×10^{15} cm^{-3}). For this study, both surfaces of the sample were passivated by a SiN_x layer. The composite of these data are shown in Fig. 14.4. The authors measured the

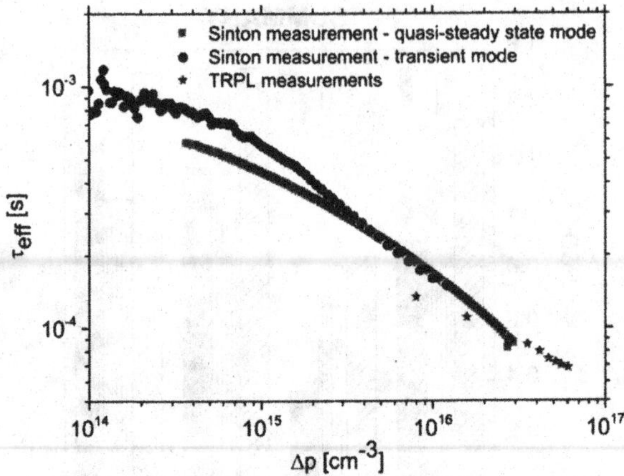

Fig. 14.4. The data of Parola and coworkers. The figure compares the lifetime measurement of a silicon wafer measured by QSSPC and the transient version of the latter. Also, TCSPC lifetime in the high injection region of the same wafer.

lifetime using QSSPC in the standard mode, and then re-measured in the transient mode. The transient mode is a built-in feature of the QSSPC apparatus. In this mode, there is an abrupt and short excitation source, followed by the capture of the photoconductive decay on a transient digitizer.

Parola also performed TCSPC using a 1.0 KHz repetition rate light source. The photon counting electronics is employed to monitor the photon counts versus time after the termination of the light pulse in the "square wave" excitation. TCSPC were reported at injection levels exceeding $1 \times 10^{16}\,\mathrm{cm}^{-3}$ with this sample and technique. The data produced about 20 to 30 percent larger lifetimes in the transient mode than for QSSPC for Δn less than about $1 \times 10^{15}\,\mathrm{cm}^{-3}$. The limited TRPL lifetime data produces lower lifetimes than the data obtained by photoconductive methods. Chapter 7 described the radiative intensity decays in the bimolecular region. We will show later in this chapter, that the radiative lifetime in high injection varies as $1/\Delta n^2$ and is one-half of the recombination lifetime measured PCD methods. The data of Parola *et al.* show the variation of the three methods over a limited range of injection level, and TRPL lifetimes are shorter in a limited range.

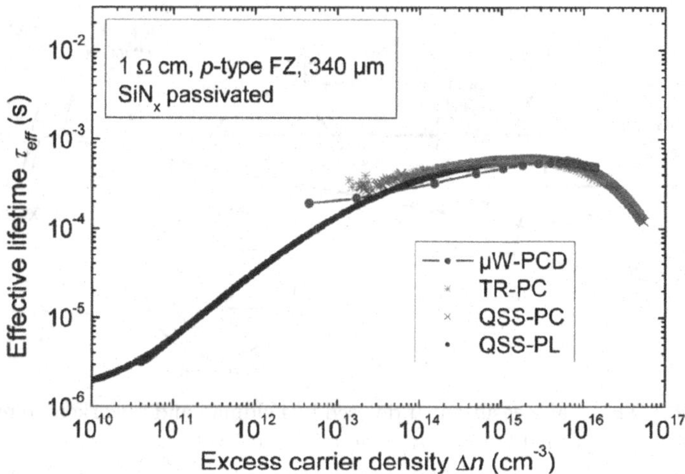

Fig. 14.5. The data or Roth and coworkers arising from the measurement of a float zone wafer by the four techniques listed. The μPCD measurement was made with a background DC light bias.

Roth and coworkers[3] performed a comprehensive measurement evaluation of silicon wafers by four different techniques. They compared μPCD, QSSPL, QSSPC and the later apparatus operated in the transient mode that they labeled TR-PC. The data displayed here was measured on a 1 ohm-cm, p-type float zone wafer that was passivated with SiN_x. High injection was attained by a complimentary light bias with the pulsed μPCD source.

14.4. Mobility Variation with Injection Level

Feldman and coworkers[4,5] devised the instrumental setup shown in Fig. 14.6. This setup allowed the sequential measurement of RCPCD and transient FCA using the same light source that was switched by means of a movable mirror. Here the injection level could be reproduced in the two measurements such that a constant level could be assumed. The same sample was moved from the FCA stage to the RCPCD stage and the transient response was measured at each station. The photoconductive transient was measured by RCPCD and the free-carrier transient was measured using a cw-CO_2 laser as

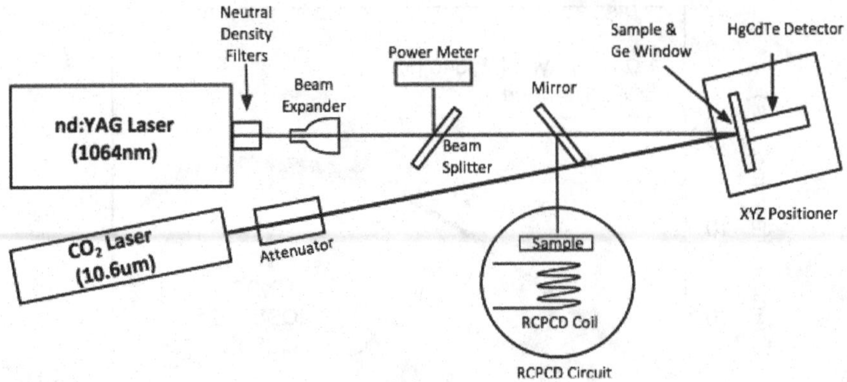

Fig. 14.6. The dual measurement devised by Feldman and coworkers to measure RCPCD and transient FCA at the same injection level.

a probe. The probe detector was an MgCdTe infrared diode. Free carrier transients, measured with this same apparatus, are shown in Fig. 11.5. The sample in that data was a float zone silicon wafer and the transient FCA was shown for a range of injection levels.

The photoconductivity resulting from electron-hole generation can be written

$$\begin{aligned}\Delta\sigma(t) &= q\Delta n(t)\mu_n + q\Delta p(t)\mu_p \\ &= q\mu_n\Delta n(t)(1+1/b).\end{aligned} \tag{14.7}$$

Here, b is the mobility ratio μ_n/μ_p. The transient photoconductivity, $\Delta\sigma(t)$, is measured by RCPCD.

The free carrier absorption can be written as:

$$\alpha = \frac{q^3\lambda^2}{4\pi^2\varepsilon_0 c n_{op}}\left(\frac{n}{(m_n^*)^2\mu_n} + \frac{p}{(m_p^*)^2\mu_p}\right). \tag{14.8}$$

For a p-type semiconductor and injection level Δn, the increase in free-carrier absorption can be written

$$\Delta\alpha(t) = \frac{q^3\lambda^2}{4\pi^2\varepsilon_0 c n_{op}\mu_n(m_n^*)^2}\left(1 + b\frac{(m_n^*)^2}{(m_p^*)^2}\right)\Delta n(t). \tag{14.9}$$

Multiplying the contemporaneous data from the two sensors, the mobility cancels:

$$\Delta\alpha(t)\Delta\sigma(t) = \frac{q^4\lambda^2(1+1/b)}{4\pi^2\varepsilon_0 c n_{op}\mu_n (m_n^*)^2}\left(1+b\frac{(m_n^*)^2}{(m_p^*)^2}\right)\Delta n^2(t). \quad (14.10)$$

One can write the result as:

$$\begin{aligned}\Delta\alpha(t)\Delta\sigma(t) &= K^2 * \Delta n^2(t);\\ \Delta n(t) &= K_2\sqrt{\Delta\alpha(t)\Delta\sigma(t)}.\end{aligned} \quad (14.11)$$

In Fig. 14.7, the RCPCD and transient FCA were measured on a p-type silicon wafer.[6] The data are co-plotted, and the time behaviors of the two measurements are radically different. The wafer under test was crystalline silicon grown by the CZ technique and had no surface passivation. This boron-doped wafer had a hole concentration of $\sim 2 \times 10^{15}\,\text{cm}^{-3}$. The RCPCD signal rises slowly from $t = 0$ to an unexpected peak at about $10\,\mu$s after the initial 5-ns excitation pulse.

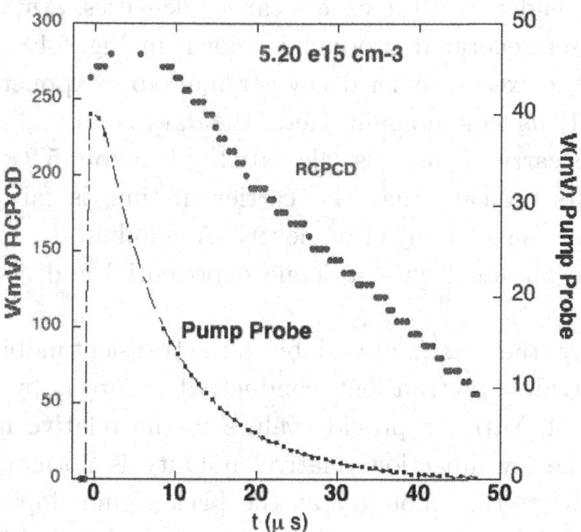

Fig. 14.7. The RCPCD and FCA transient data for a 10 ohm-cm silicon wafer using an initial injection level of $6.20 \times 10^{15}\,\text{cm}^{-3}$.

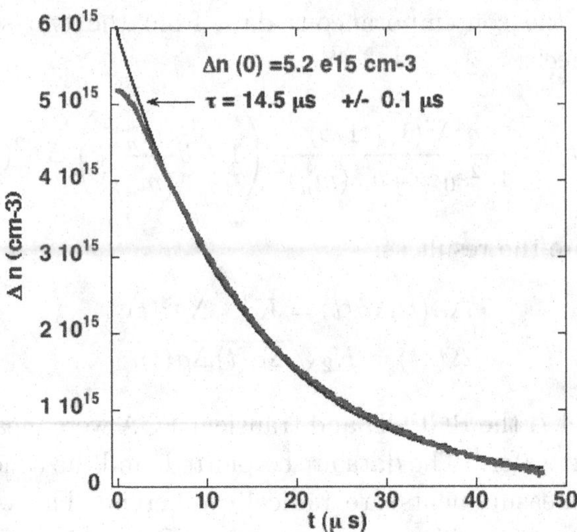

Fig. 14.8. The calculated value of the excess electron concentration as a function of time after an initial injection of 5.2×10^{15} cm^{-3}.

Combining the RCPCD and pump-probe raw data and multiplying the two data sets, the resulting square root of the product contain the time dependence of the excess carrier densities, $\Delta n(t)$ or $\Delta p(t)$. The computer-generated product is shown in Fig. 14.8. A data fit shows that the excess carrier decay is remarkably exponential except in the 0 to 10 μs time domain. Here, the data is normalized but the initial excess carrier density is calculated to be about 5.2×10^{15} cm^{-3}. These results indicate that the carrier lifetime is fairly constant over a wide range of injection levels. A good fit to the product data can be obtained with a single exponential and a decay time of 14.5 μs.

Therefore, the dominant variable is the transient mobility, $\Delta\mu(t)$. One can divide the transient conductivity, $\Delta\sigma(t)$, by the calculated value of $\Delta n(t)$ to provide values of the relative mobility. In Fig. 14.9, the low injection relative mobility is normalized to the theoretical scattering mobility at the background doping level, or 2×10^{15} cm^{-3}. From the tables, that value is found to be 1200 cm^2/Vs. The normalized mobility is seen to decrease about a factor of six from that the highest injection level or about 200 cm^2/Vs. The RCPCD

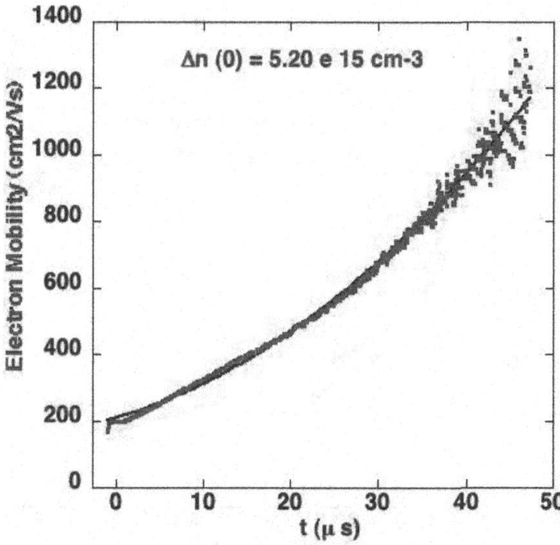

Fig. 14.9. The electron mobility calculated by combining the RCPCD and FCA transient data. [© (2019) IEEE. Reprinted with permission, from Ahrenkiel, R. K., Feldman, A., Lehman, J., and Johnston, S. W. *IEEE J. Photovoltaics* **3**, pp. 348–352 (2013).]

peak results from the observation that the mobility increases faster than the excess carrier decreases in this injection range.

The calculated ambipolar mobility is plotted in Fig. 14.10 and combined with data shown in Fig. 14.9. The plots show that the ambipolar model fits the data very well. In summary, the effects shown here are in good agreement with the ambipolar transport theory and account quite well for the behavior of both PC decay and free carrier decay.

As a result of these studies, one must conclude that mobility changes, due to the space charge effects and ambipolar diffusion, are a significant variable in the high-injection regime of photoconductive decay. These changes may be much larger than the increases in carrier scattering rate that decrease the mobility at high injection.

One can describe the conductivity peak in Fig. 14.7, by allowing for the variation of mobility with excess carrier density as:

$$\frac{d\Delta\sigma(t)}{dt} = q\left(\mu\frac{d\Delta n(t)}{dt} + \Delta n(t)\frac{d\mu(t)}{dt}\right). \tag{14.12}$$

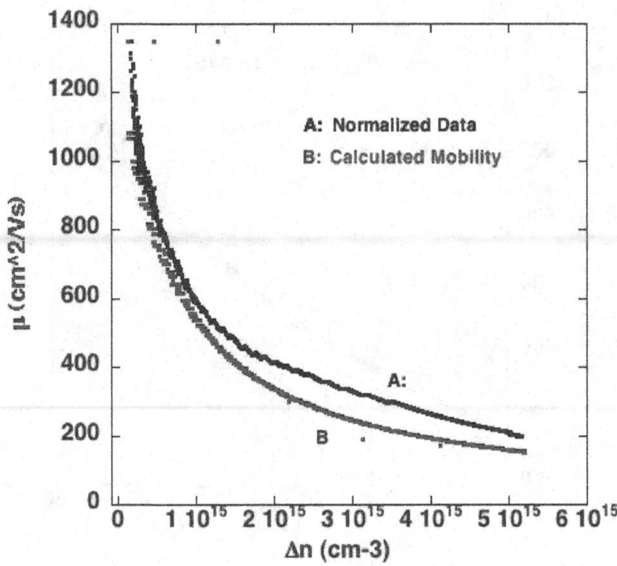

Fig. 14.10. The normalized electron mobility is calculated from the RCPCD-FCA data and shown as curve A. The theoretical mobility is calculated from ambipolar theory and uses the injection levels of $5.2 \times 10^{15}\,\text{cm}^{-3}$ (curve B).

The first term is inherently negative, and the second term is positive.

$$\left|\Delta n(t)\frac{d\mu(t)}{dt}\right| - \left|\mu\frac{d\Delta n(t)}{dt}\right| > 0. \tag{14.13}$$

When the conductivity will undergo a peak as seen in Fig. 14.7 when the two terms are equal. Thus, PCD data analysis should include ambipolar effects to accurately extract carrier lifetime.

14.5. Simultaneous Measurements

Recent work combined PCD and TRPL into a single measurement that provides simultaneous acquisition of photoconductive decay and time-resolved photoluminescence data.[8] The PCD signal is obtained by a new method labeled transmission-modulated photoconductive decay (TMPCD)[9] method. The basic TMPCD apparatus is complemented by a detector to records the simultaneous photoluminescence generated by the sample.

A schematic of the apparatus used to make transient PCD and TRPL measurements is shown in Fig. 14.11. The sample bridges the transmitting and receiving coils, which couple the 500 MHz electromagnetic wave (EMW) from the transmitter to the receiver. A portion of the transmitted wave penetrates the sample and is "trapped" by total internal reflection. As such, the wave travels along the lateral direction of the sample, and a portion of that wave radiates from the bottom surface to the receiving antenna.

The absorption coefficient of the EMW in a semiconducting medium has been calculated[9] and is:

$$\alpha = \omega\sqrt{\frac{\mu\varepsilon}{2}\left(\sqrt{1+\frac{\sigma^2}{\omega^2\varepsilon^2}}-1\right)} \tag{14.14}$$

Where α is the attenuation coefficient of the electric vector in the medium and σ is the conductivity:

$$E = E_0 \exp\left(-\alpha x/2\right) \exp\left[i(\omega t - \beta x)\right]. \tag{14.15}$$

When the conductivity, σ, is increased by photon absorption, the E-vector component of the EMW is attenuated. At the frequency used here ($\sim$500 MHz), the attenuation factor is fairly small, but the larger path produced by wave trapping increases the interaction length. This longer path length provides the larger signals that were observed these experiments.

As the detector is open for detection of photoluminescence, the transient photoluminescence decay data (TRPL) is simultaneously collected by a photodetector. The PCD data and the analogue TRPL data are collected on a dual-channel digitizer. The combined, simultaneous analog data acquisition provides a unique signature that allows analysis of the transient photoluminescence (TRPL) and PCD decay. The dual system setup is shown in Fig. 14.11.

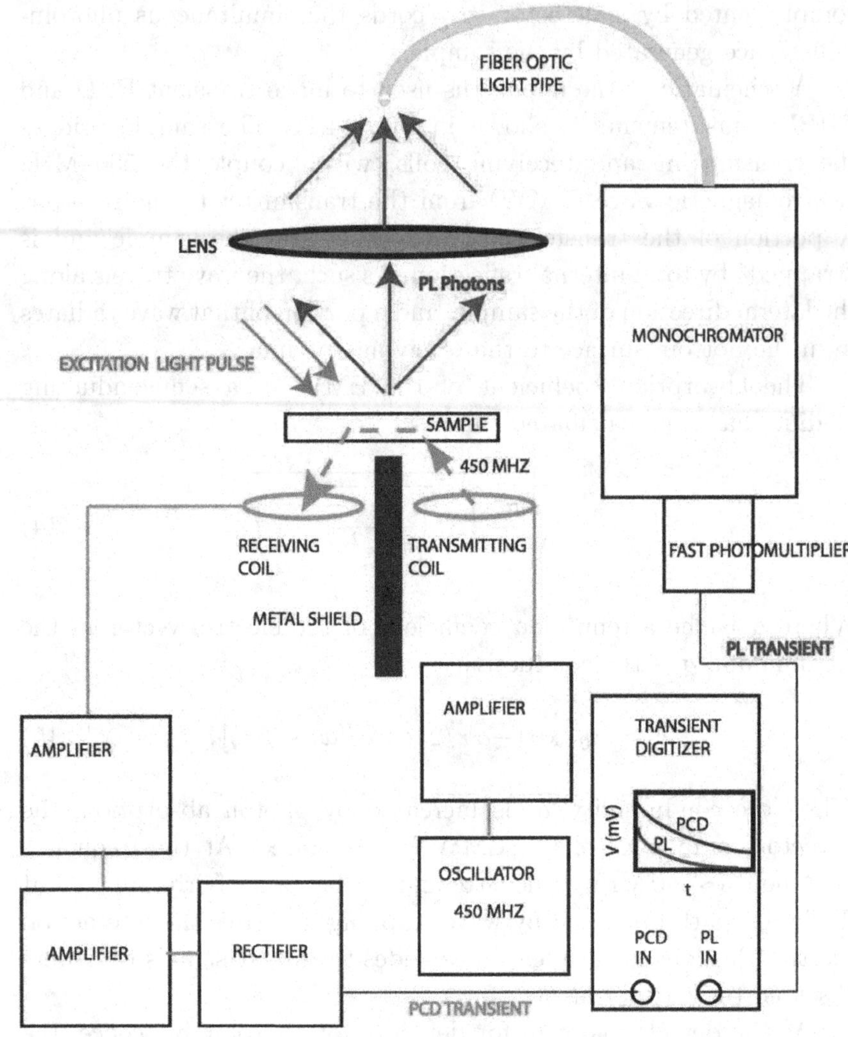

Fig. 14.11. Schematic of dual TMPCD/TRPL apparatus. [© (2019) IEEE. Reprinted with permission, from Ahrenkiel, R. K., Feldman, A., Lehman, J., and Johnston, S. W. *IEEE J. Photovoltaics* **3**, pp. 348–352 (2013).]

The data in Fig. 14.12 was measured on a GaAs/GaInP double heterostructure (DH) thin film grown by metal-organic chemical vapor deposition (MOCVD). The sample was not intentionally doped, but the background doping is in the $7 \times 10^{15}\,\mathrm{cm}^{-3}$ range

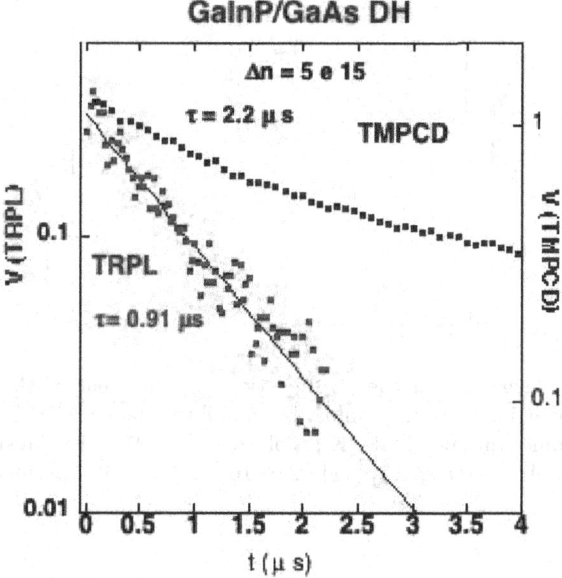

Fig. 14.12. Data measured on a GaInP/GaAs DH using the apparatus of Fig. 14.11. [© (2019) IEEE. Reprinted with permission, from Ahrenkiel, R. K., Feldman, A., Lehman, J., and Johnston, S. W. *IEEE J. Photovoltaics* **3**, pp. 348–352 (2013).]

and the thickness is 2 μm. The lowest-injection data of the GaAs DH are shown in Fig. 14.12. These data show that the initial decay times vary by about a factor of two. The injection level here is estimated at 5×10^{15} cm^{-3}. Using the equation (14.5), the lifetime ratio is 1.3 with no mobility correction. A numerical simulation of this experimental results is plotted in Fig. 14.13. The plot of this calculation shows the time variation of the PL decay, the excess carrier decay, and the excess conductivity decay. The behavior is similar to that measured by the dual sensor apparatus.

A measurement that accentuates the ambipolar mobility change is shown in Fig. 14.14. The initial injection level is increased to 2.5×10^{16} cm^{-3}. The mobility variation clearly produces a longer effective decay time and produces a larger variation between PCD and TRPL data. The TRPL decay decreases and the PCD decay time increases as the theory predicts.

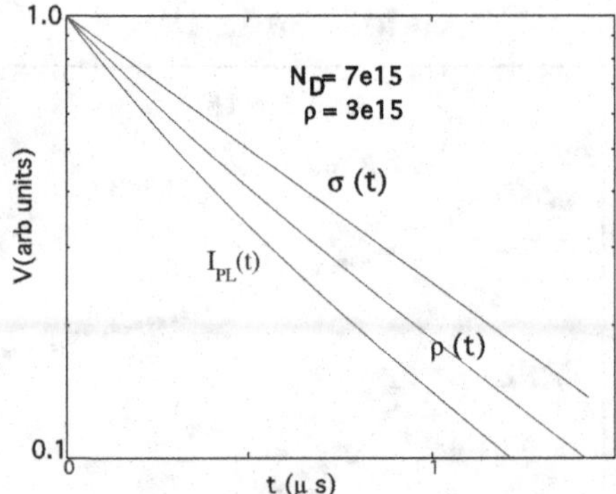

Fig. 14.13. Calculation of the low-injection measurement of GaAs by TRPL and PCD, which includes the ambipolar mobility change with injection level. [Reprinted from Ahrenkiel, R. K., Johnston, S. W., Kuciauskas, D. and Tynan, J., *J. Appl. Phys.* **116**(21), 214510 (2014), with permission of AIP Publishing.]

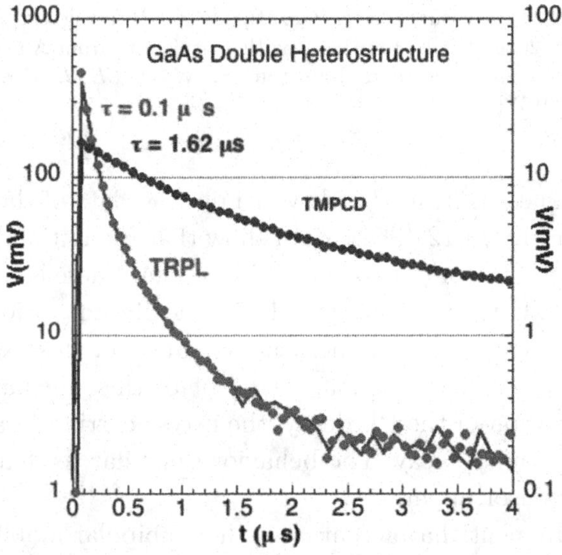

Fig. 14.14. Simultaneous measurement of PCD and TRPL on the GaAs DH with an initial injection level of 2.5×10^{16} cm^{-3}. [Reprinted from Ahrenkiel, R. K., Johnston, S. W., Kuciauskas, D. and Tynan, J., *J. Appl. Phys.* **116**(21), 214510 (2014), with permission of AIP Publishing.]

14.6. Triple Sensor Measurements

S. W. Johnston[10] and coworkers developed a very unique triple sensor apparatus that measures carrier lifetime simultaneously by TRPL, μPCD, and TFCA (transient FCA). This apparatus produced lifetime data using the same excitation source. This situation insures that the carrier injection level and profile is identical for all three measurements. A schematic of the configuration is shown in Fig. 14.15. A Nikon camera lens collects the photoluminescence from the illuminated surface and focused the collected light onto input slit of a spectrometer. The light collection apparatus is tuned to the band-to-band luminescence of the particular material. A photomultiplier converts the PL decay signal to channel 1 of a three channel transient digitizer. This is basically an analogue TRPL system described in Chapter 7. The pulsed laser system is

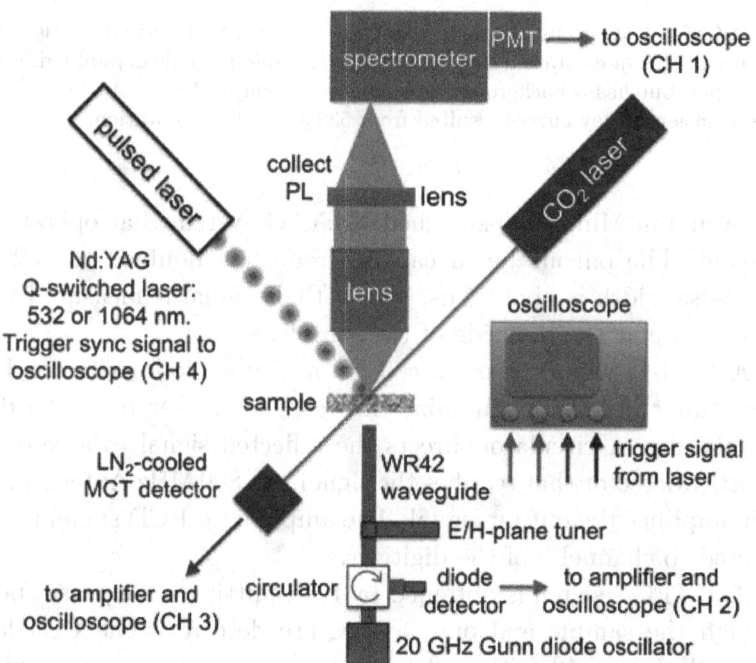

Fig. 14.15. Schematic of the simultaneous three transient technique of S. W. Johnston and coworkers.

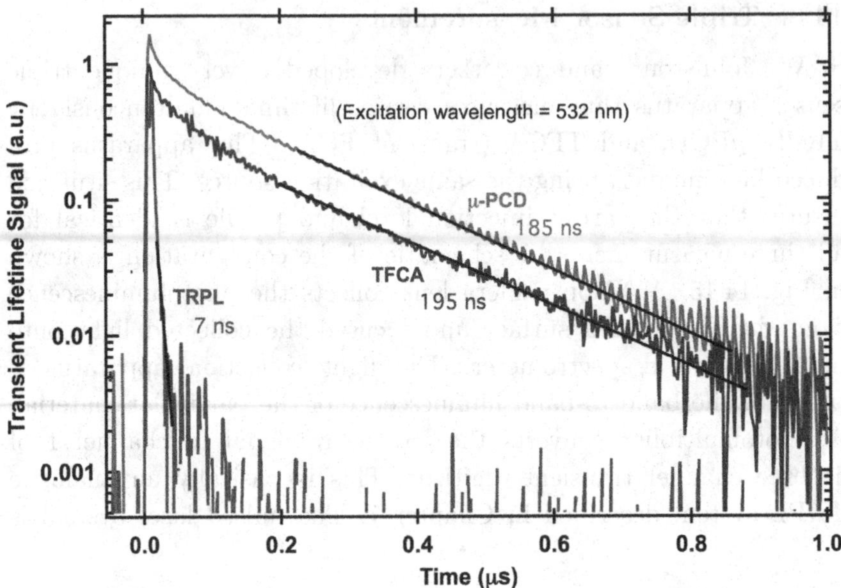

Fig. 14.16. Data from the three transient measurement system of Fig. 14.15 that use a 532 nm excitation wavelength. The sample is single-crystal CdTe that is undoped but has a background hole concentration of $2 - 3 \times 10^{14}$ cm^{-3}. The three transient decay curves resulted from 532 nm pulsed excitation.

a Continuum Minilite Q-switched Nd:YAG system that operates at 1064 nm. The output beam can be frequency doubled to 532 nm. The pulse width is about 5 ns. The μ PCD signal is measured from a waveguide at the backside of the sample.

A 20 GHz signal is produced by a Gunn diode with an E/H plane tuner in-line to minimize the reflected power from the dark conductivity. A circulator directs the reflected signal to a zero-bias Schottky detector that rectifies the signal. An 800 MHz Avtech 141C1 then amplifies the output signal. The amplified μ PCD signal is then directed to channel 2 of the digitizer.

The TFCA signal is obtained by transmitting a CO_2 laser beam through the sample and onto a HgCdTe detector. The CO_2 laser is a 1 W LASY-48 with a 2.4 mm beam size. The wave guide is retracted from the back surface by several mm to allow the CO_2 beam to pass through the sample at a 45 degree angle onto the HgCdTe detector. This signal is also amplified by a wide-bandwidth amplifier and connected to channel 3 of the digitizer.

The nominally undoped sample is single crystal CdTe grown by the vertical gradient freeze technique. The crystal dimensions are 0.8 mm thick by 1 cm × 1 cm cross section. The background doping shows *p*-type conductivity with a hole concentration of $2–3 \times 10^{14}\,\text{cm}^{-3}$. The surfaces are unpassivated and so the intrinsic SRV is expected to be quite large. The data of Fig. 14.13 shows the set of room-temperature decay curves with an injection level of $\sim 5 \times 10^{15}\,\text{cm}^{-3}$ after excitation by a 532 nm YAG laser pulse. The pulse width is 5 ns and the repetition rate is 15 Hz. The TRPL data shows a fast decay that is basically system response whereas the transient free carrier (TFCA) and microwave reflection μ-PCD show decay times of 185–195 ns. The band pass of the detection monochromator is tuned the 850 nm peak emission to pass to the photomultiplier. The absorption coefficient of CdTe at 532 nm is about $1 \times 10^5\,\text{cm}^{-1}$ so the Beers law absorption depth is about $0.1\,\mu$m. The PL signal is only observable when the excess carriers are near the entrance surface whereas the TFCA and μ-PCD probe the entire sample. The fast decay of the PL signal reflects that the SRV of the front surface dominates the TRPL recombination. Using the electron lifetime of 190 ns and the listed electron mobility of CdTe as $600\,\text{cm}^2/\text{Vs}$, the electron diffusion length here is $17\,\mu$m and is much longer than absorption length of $0.1\,\mu$m. Thus, a significant fraction of emitted photons are self-absorbed after electron diffusion into the bulk. These photons can either generate another photon or phonons by the SRH process. Clearly, a significant number of secondary photons are not generated by self-absorption.

Photon recycling is negligible if the bulk SRH recombination rate is much larger than the radiative recombination rate: i.e.

$$\frac{1}{\tau_{\text{NR}}} \gg \frac{1}{\tau_{\text{R}}}: \tag{14.16}$$

In this case, the self-absorbed, deeply generated photons, have a high probability or decaying by the SRH process. The only photons that are externally emitted are those on the low-energy (long wavelength) side of the emission band that are transmitted without absorption.

As the direct absorption spectra of CdTe is similar to that of GaAs. Using the B-coefficient of GaAs, the calculated radiative lifetime is 2.5 μs. The inequality of Eq. (14.17) clearly applies here and the internal quantum efficiency is approximately 0.07. Thus photon regeneration of diffused electrons is a weak process here. The majority of self-absorbed photons decay by phonon generation. However, a significant concentration of photo-generated electrons are generated in the bulk and are detected by TFCA and μ-PCD.

A second measurement was made by Johnston and coworkers with the three sensor apparatus but using the fundamental YAG wavelength (1064 nm) as the excitation source. The CdTe is transparent to 1064 nm in first order. At high excitation intensity, the second order absorption process produces a weak but uniform absorption in the volume of the CdTe crystal. The transient data are shown in Fig. 14.17 and there is a PL decay signal with a 160 ns decay time.

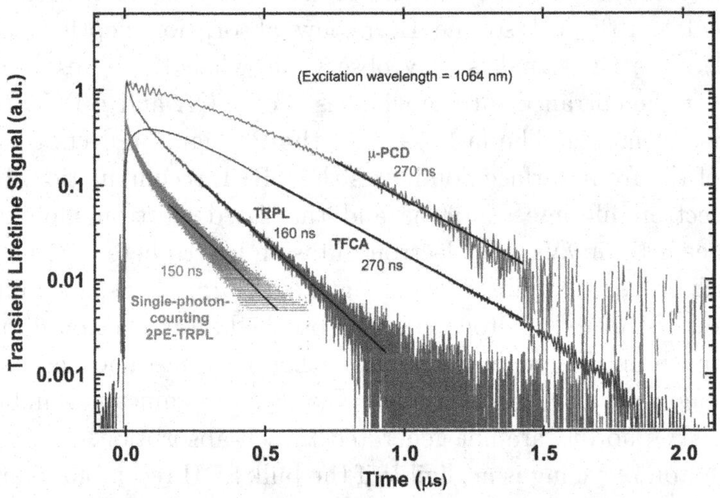

Fig. 14.17. The triple sensor data on the same CdTe crystal using the YAG 1024 nm wavelength to create electron-hole pairs uniformly in the volume by means of second harmonic generation. In this case, the TRPL signal is detectable with this system. Also shown for comparison is the time-correlated single photon measurement using that dedicated apparatus. [© (2019) IEEE. Reprinted with permission, Johnston, S., Zaunbrecher, Ahrenkiel, R. K., Kuciauskas, D., Albin, D., and Metzger, W. Simultaneous measurement of minority-carrier lifetime in single-crystal CdTe using three transient decay techniques, *IEEE J. Photovoltaics* 4, 1295–1300 (2014).]

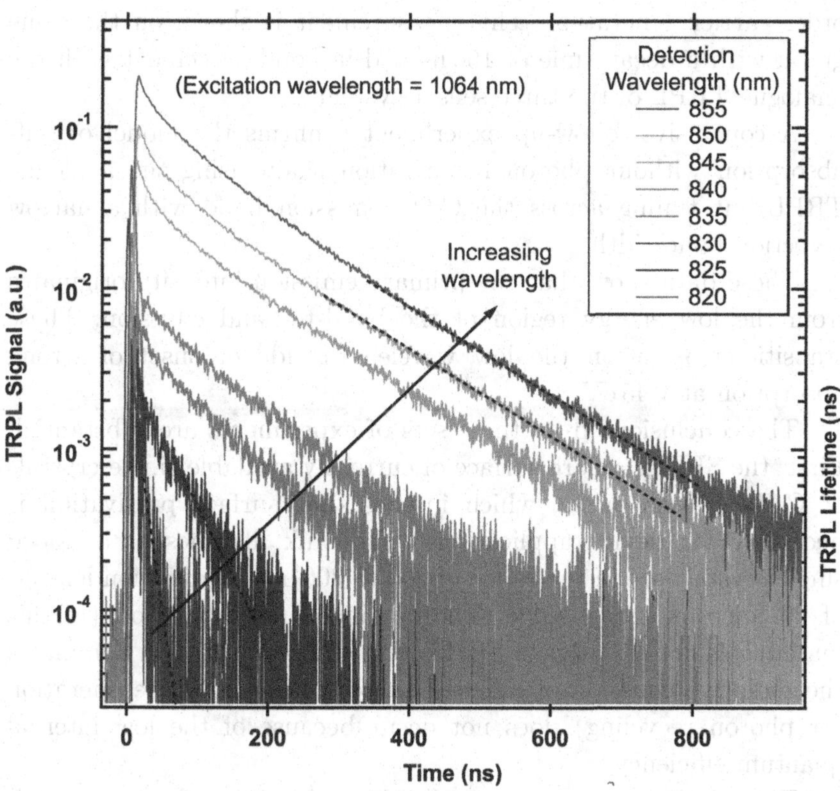

Fig. 14.18. The TRPL decay of the CdTe crystal using second harmonic generation (1064 nm) and tuning across the emission band from 820 nm (1.51 eV) to 855 nm (1.45 eV). [© (2019) IEEE. Reprinted with permission, Johnston, S., Zaunbrecher, Ahrenkiel, R. K., Kuciauskas, D., Albin, D., and Metzger, W. Simultaneous measurement of minority-carrier lifetime in single-crystal CdTe using three transient decay techniques, *IEEE J. Photovoltaics* **4**, 1295–1300 (2014).]

The TFCA and μ-PCD data show somewhat longer decay times of 270 ns each. The increase of the latter lifetimes are related to more carriers being isolated from the surface recombination effects. The emergence of an intrinsic PL signal from the band edge supports the model that only the low energy, primary photons are able to be externally emitted and the higher energy photons are self-absorbed and converted to phonons. An independent measurement was made by a time-correlated single photon counting systems, using second order carrier generation. This measurement is shown on the same graph with a decay time of 150 ns and in good agreement with the analogue TRPL of the three sensor system.

A conclusive follow-up experiment confirms the model of self-absorption without photon regeneration again using the analogue TRPL but tuning across the CdTe emission band with a narrow detection bandwidth.

These data show that the primary emission intensity originates from the low energy region of the band-to-band emission. These transitions occur on the low wavelength side of onset of strong absorption at 1.45 eV.

The conclusions from these sets of experiments are substantial. First, the SRV of a bare surface of currently available single crystals of CdTe is very large, which implies that surface passivation is necessary for device applications. The bulk lifetimes of undoped single crystals are in the range of 200 to 300 ns and diffusion lengths of 20 microns are possible. The dominant recombination (in this material) is nonradiative or SRH in origin. Self-absorption dominates the radiative emission of diffused carriers, but photon regeneration (or photon recycling) does not occur because of the low internal quantum efficiency.

The results on single crystal CdTe adds some further considerations regarding the photon recycling discussions of Chapter 13. Self-absorption is an inherent property of direct bandgap materials because of the universal applicability of the van Roosbroeck–Shockley relationship. However, photon recycling only occurs when the internal quantum efficiency is near unity.

14.7. Summary

The comparison of lifetime measurements by either sequential or simultaneous measurement indicates that the results are technique dependent. One unique case is the measurement of photoconductive and photoluminesce lifetime at low injection in double confinement structures. Data in Chapter 12 show an example of that agreement for a GaAs DH, and that result agrees with the theory.

Other results shown in previous chapters show that TRPL and PCD have a different functional dependence at high injection. The transient decay function can be analyzed to provide the dominant carrier decay mechanism. Each mechanism has a unique carrier decay behavior that is manifested in decay behavior. We have presented such behavior when the dominant mechanism is radiative recombination, SRH dominated recombination, Auger recombination, or shallow trapping. In addition, measurements have produced unique decay properties of charge separation structures. Finally, the variation of mobility with injection level has a prominent role photoconductive-based measurements. There are likely to be other examples that demonstrate the utility of transient measurements in material diagnostics.

References

1. Ahrenkiel, R. K., Call, N., Johnston, S. W., and Metzger, W. K. *Solar Energy Materials and Solar Cells* **95** (2010), 2197–2204.
2. Stéphanie Parola, Mehdi Daanoune, Alexandru Focsa, Bachir Semmache, Erwann Picard, Anne Kaminski-Cachopo, Mustapha Lemiti, Danièle Blanc-Pélissier, *Energy Procedia* **55** (2014) 121–127
3. Roth T, Rosenits P, Rudiger M, Warta W, and Glunz SW. 2008. In: 2008 Conference on Optoelectronic and Microelectronic Materials and Devices. pp. 249–252.
4. Ari Feldman, Richard Ahrenkiel, and John Lehman, "Transient mobility in silicon as seen by a combination of free-carrier absorption and resonance-coupled photoconductive decay", *J. Appl. Phys.* **112**, 103703 (2013).
5. Ari Feldman, Richard Ahrenkiel, and John Lehman, "Degradation of photovoltaic devices at high concentration by space charge limited currents", *Solar Energy Materials and Solar Cells* **117**, pp. 408–411 (2013).
6. Ahrenkiel, R. K., Feldman, A., Lehman, J., and Johnston, S. W. *IEEE J. Photovoltaics* **3**, pp. 348–352 (2013).
7. Ahrenkiel, R. K., Johnston, S. W., Kuciauskas, D. and Tynan, J. *J. Appl. Phys.* **116**(21), 214510 (2014).
8. Ahrenkiel, R. K. and Dunlavy, D. J. "A new lifetime diagnostic system for photovoltaic materials," *Solar Energy Materials and Solar Cells* **95**, 1989 (2011).

9. Ahrenkiel, R. K. and S. W. Johnston, "The interaction of microwaves with photoelectrons in semiconductors", *J. Vac. Sci. Technol. B.* **26** (2008).
10. Steve Johnston, Katherine Zaunbrecher, Richard Ahrenkiel, Darius Kuciauskas, David Albin, Wyatt Metzger, "Simultaneous measurement of minority-carrier lifetime in single-crystal CdTe using three transient decay techniques", *IEEE Journal of Photovoltaics* **4**, 1295–1300 (2014).

CHAPTER 15

Summary and Future Work

The work described in this manuscript has been focused on the electrical and optical properties of electronic and photovoltaic materials. In particular, the work has focused on measurement of carrier lifetime by new and improved methods. The first author wrote a chapter on carrier lifetime measurements in *Minority Carriers In III-V Semiconductors: Physics and Applications* series[1] in 1993. That topic has proven to be very popular based on the continuing demand for reprints. This work is based on updating and expanding that original content to include most of the materials of current interest to the photovoltaic community. The low dimensional materials such as quantum dots and wires are not included in this work. The new, emerging materials based on the perovskite crystalline structure, are also excluded in this writing. These areas will be addressed in future work.

Areas of considerable interest, in commercial photovoltiacs, are rapid measurements that are compatible with the production environment. Fast measurements, that are applicable to early stages of the production process, are desirable for lower-cost, quality control of commercial devices. Mapping techniques are very valuable for the polycrystalline materials as spatial uniformity is almost always a primary issue in these materials.

This work does not include the vast infrastructure of methods to determine materials composition. These include the vacuum technologies that provide material and structural information. Some current work in the laboratory involves combining electron microscopy with luminescence properties. This allows instant correlation of electronic properties with material composition and chemistry. Some of the aforementioned work will be included in any future manuscripts that follow in this series.

Technique development of characterization tools has impacted all of the major technologies, from semiconductor to biomedical. Prior to the development of the magnetic resonance imaging, computer aided tomography (CAT) scans, and ultrasonic imaging, exploratory surgery was a standard for the diagnosis of illness. At the same time, contactless, nondestructive measurement has become a standard in the electronics industry. Continued investment in technique development in all of these areas will lead to improvements in the next generation of technology.

Reference

1. R. K. Ahrenkiel, *Minority Carriers in III-V Semiconductors, Physics and Applications*, V. 39, pp. 40–145, Academic Press, Inc., 1993.

Index

World Scientific Series in Materials and Energy

(Continuation of series card page)

Vol. 12 *World Scientific Handbook of Organic Optoelectronic Devices (Volumes 1 & 2)*
edited by Franky So (North Carolina State University, USA)

Vol. 11 *Industrial Applications of Ultrafast Lasers*
by Richard Haight (IBM TJ Watson Research Center, USA) and Adra V. Carr (IBM TJ Watson Research Center, USA)

Vol. 10 *World Scientific Reference on Spin in Organics (In 4 Volumes)*
edited by Zeev Valy Vardeny (University of Utah, USA) and Markus Wohlgenannt (University of Iowa, USA)

Vol. 9 *Conjugated Polymers and Oligomers: Structural and Soft Matter Aspects*
edited by Matti Knaapila (Technical University of Denmark, Denmark)

Vol. 8 *Thin Films on Silicon: Electronic and Photonic Applications*
edited by Vijay Narayanan (IBM Thomas J Watson Research Center, USA), Martin M. Frank (IBM Thomas J Watson Research Center, USA) and Alexander A. Demkov (The University of Texas at Austin, USA)

Vol. 7 *The WSPC Reference on Organic Electronics: Organic Semiconductors (In 2 Volumes)*
edited by Jean-Luc Brédas (King Abdullah University of Science & Technology, Saudi Arabia & Georgia Institute of Technology, USA) and Seth R. Marder (Georgia Institute of Technology, USA)

Vol. 6 *Handbook of Solid State Batteries (Second Edition)*
edited by Nancy J. Dudney (Oak Ridge National Laboratory, USA), William C. West (Nagoya University, Japan) and Jagjit Nanda (Oak Ridge National Laboratory, USA)

Vol. 5 *Handbook of Green Materials: Processing Technologies, Properties and Applications (In 4 Volumes)*
edited by Kristiina Oksman (Luleå University of Technology, Sweden), Aji P. Mathew (Luleå University of Technology, Sweden), Alexander Bismarck (Vienna University of Technology, Austria), Orlando Rojas (North Carolina State University, USA), and Mohini Sain (University of Toronto, Canada)

Vol. 4 *Materials for Biofuels*
edited by Arthur J. Ragauskas (Georgia Institute of Technology, USA)

Vol. 3 *The World Scientific Handbook of Energy*
edited by Gerard M. Crawley (University of South Carolina, USA)